# Youth and agriculture:

## Key challenges and concrete solutions

Published by the Food and Agriculture Organization of the United Nations (FAO) in collaboration with the Technical Centre for Agricultural and Rural Cooperation (CTA) and the International Fund for Agricultural Development (IFAD)

(FAO) ISBN 978-92-5-108475-5
(FAO) E-ISBN 978-92-5-108476-2 (PDF)
(CTA) ISBN 978-92-9081-558-7

The resources used by FAO for this document have been provided by the Swedish International Development Cooperation Agency (Sida). Sida does not necessarily share the views expressed in this material. Responsibility for its content rests entirely with the authors.

Cover photo: © IFAD/Susan Beccio

# Contents

Foreword                                                                    v
Acknowledgments                                                            vii
Executive summary                                                           ix
Acronyms                                                                     xi
Introduction                                                               xvii

**1. Access to knowledge, information and education**                        **1**

  1.1 Introduction                                                  2

  1.2 Case studies                                                  4

     1.   Agri-enterprise development and management    4

     2.   Rebranding agriculture in schools             5

     3.   Young Women Open Schools                       6

     4.   On-the-job training                           7

     5.   PhD training in agriculture                   9

     6.   Distance learning for young farmers           10

     7.   ICTs for extension services                   11

     8.   ICT solutions for agriculture                 13

     9.   Youth resource centres on agriculture         15

  1.3 Conclusions                                                   16

**2. Access to land**                                                       **19**

  2.1 Introduction                                                  20

  2.2 Case studies                                                  21

     10.   Land tenure, farm productivity and enterprise development    21

     11.   Land ownership for shea butter producers    23

     12.   Distributing hillside land to landless youth    24

     13.   Young rural entrepreneur and land fund programme    24

     14.   Reclaiming desert land for young graduates    26

     15.   Small landlords and large tenants programme    28

     16.   Short-term land leases for youth              28

  2.3 Conclusions                                                   30

**3. Access to financial services**                                         **33**

  3.1 Introduction                                                  34

  3.2 Case studies                                                  35

     17.   Installation aid                             35

     18.   Public-private investment fund               37

     19.   Youth Venture Capital Fund                   38

     20.   Youth socio-economic empowerment service     38

     21.   Financial services for youth through rural entrepreneurship    39

     22.   Friends Help Friends Saving Group            40

     23.   Loan project for young entrepreneurs         42

     24.   Crowd-funding: The goat dairy project        43

     25.   Finance and mentorship for innovative young social entrepreneurs    44

  3.3 Conclusions                                                   45

## 4. Access to green jobs     49
4.1 Introduction     50
4.2 Case studies     51
    26.    Junior Farmer Field and Life School programme     51
    27.    Vocational training in small biogas companies     52
    28.    Vocational training for young beekeepers     53
    29.    Green jobs apprenticeship programme     54
    30.    Raising youth's awareness of organic agriculture     55
    31.    Agro-ecotourism business     57
    32.    Transforming water hyacinth into paper     59
4.3 Conclusions     60

## 5. Access to markets     63
5.1 Introduction     64
5.2 Case studies     65
    33.    Connecting farmers     66
    34.    Innovation in distribution and sales     66
    35.    Linking producers and consumers     68
    36.    Certifying social youth business     69
    37.    Milk and dairy processing     70
    38.    Innovative models for young coffee producers     71
    39.    Independent business born from the commitment of youth     72
5.3 Conclusions     73

## 6. Engagement in policy dialogue     77
6.1 Introduction     78
6.2 Case studies     79
    40.    Young farmers' representation     79
    41.    The African Union listening to youth     81
    42.    Documenting youth policies and initiatives     82
    43.    Youth peasant federation     84
    44.    Echoing the voices of Youth     84
    45.    Youth caring about the environment     85
    46.    The European Council of Young Farmers     87
    47.    YPARD – Young Professionals' Platform for Agricultural Research
          for Development     88
6.3 Conclusions     89

## 7. Overall conclusions     91

## References     97

## Annex I: Survey 'Youth and agriculture: key challenges and concrete solutions'     101

# Foreword

Rural youth are the future of food security. Yet around the world, few young people see a future for themselves in agriculture or rural areas.

Rural youth face many hurdles in trying to earn a livelihood. Pressure on arable land is high in many parts of the world, making it difficult to start a farm. Youth often also lack access to credit, and many other productive resources necessary for agriculture. But even if such hurdles can be overcome, isn't urban life much cooler? Perhaps, but not if you cannot make a living there. Particularly in developing countries, rural youth find themselves in such a bind.

While most of the world's food is produced by (ageing) smallholder farmers in developing countries, older farmers are less likely to adopt the new technologies needed to sustainably increase agricultural productivity, and ultimately feed the growing world population while protecting the environment. Hence, we need to re-engage youth in agriculture. Can this be done?

This publication provides real life examples of how this can be done. It shows how tailor-made educational programmes (such as the Junior Farmer Field and Life Schools approach) can provide rural youth with the skills and insights needed to engage in farming and adopt environmentally friendly production methods. With some additional effort, through farmer organizations and improved infrastructure, young farmers can connect to markets to sell their often higher value food. Facilitating youth's access to credit helps them become entrepreneurs, improving their self-esteem and the feeling that they can make a living in rural areas.

None of this will come easily. There are no silver bullets. However, the large number of successful initiatives presented in this study offer a sense of hope. There are workable solutions to overcome the challenges faced by young women and men trying to engage in agriculture and earn a living in rural areas. Many of the initiatives reported in this study originate with the youth themselves. These initiatives show that – when there is a supportive environment – youth are able to find innovative ways to create a future for themselves, and also contribute to the societies and communities in which they live.

We hope that this publication will help development practitioners, youth leaders, youth associations, producers' organizations and policy makers alike by providing insights into possible solutions that can be tailored to their own context.

This study has been a joint undertaking of the Food and Agriculture Organization of the United Nations (FAO), the International Fund for Agricultural Development (IFAD) and the Technical Centre for Agricultural and Rural Cooperation (CTA). We are most grateful to our staff for their initiative and contributions in putting together this compilation of good practices, and for showing us the lessons we all should draw from those experiences.

**Rob Vos**, Director of the Social Protection Division of FAO (ESP)
**Xiangjun Yao**, Director of the Climate, Energy and Tenure Division of FAO (NRC)
**Marcela Villarreal**, Director of the Office for Communication, Partnerships and Advocacy of FAO (OPC)
**Adolfo Brizzi**, Director of Policy and Technical Advisory Division of IFAD (PTA)
**Lamon Rutten**, Manager PMI (CTA)

# Acknowledgments

The preparation of Youth and agriculture: key challenges and concrete solutions was possible thanks to the valuable contributions of individuals, divisions and organizations, and their inputs are gratefully acknowledged. The publication was prepared by staff of the Climate, Tenure and Energy Division (NRC), the Social Protection Division (ESP) and the Office for Partnerships, Advocacy and Capacity Development (OPC) of FAO; the Agriculture, Rural Development and Youth in the Information Society of CTA; and the Programme Management Department (PMD) and the Technical Advisory Division (PTA) of IFAD. Overall guidance was provided by Nora Ourabah Haddad (OPC), Bernd Seiffert (ESP), Reuben Sessa (NRC) and Rosalud de La Rosa (OPC) from FAO; Anne-Laure Roy from IFAD; and Ken Lohento from the ARDYIS project of CTA. The chapters on access to knowledge, information and education; access to land; and engagement in policy dialogue were prepared by Charlotte Goemans (FAO–OPC). The chapters on access to financial services and access to markets were prepared by Francesca Dalla Valle (FAO–ESP), Alessandra Giuliani and Martina Graf from the Bern University of Applied Sciences, School for Agricultural, Forest and Food Sciences (HAFL), and Charlotte Goemans. The chapter on access to green jobs was prepared by Tamara van 't Wout (FAO–NRC).

The following FAO colleagues provided information on FAO's youth work and/or provided comments on the publication: Olivio Argenti, Ida Christensen, Eve Crowley, Rosalud de la Rosa, Ana Paula de la O Campos, Nicoline De Haan, Eric Demafeliz, Francesca Distefano, Farid El Haffar, Valentina Franchi, Boris Gandon, Alashiya Gordes, Nandini Gunewardena, Mitchell Hall, Mirela Hasibra, Malcolm Hazelman, Susan Kaaria, Daniela Kalikoski, Regina Laub, Sharon Lee Cowan, Erdgin Mane, Raffaele Mattioli, Katia Meloni, Valeria Menza, Calvin Miller, Jamie Morrison, Shirley Mustafa, Olga Navarro, Courtney Paisley, Hajnalka Petrics, George Rapsomanikis, Mariagrazia Rocchigiani, Nicholas Ross, Makiko Taguchi, Rob Vos, Peter Wobst, Andrea Woolverton and James Zingeser.

The following IFAD colleagues provided information on IFAD's youth work and/or provided comments on the publication: Abdelhamid Abdouli, Hubert Boirard, Hawa Bousso, Amadou Daouda Dia, Ambra Gallina, Beatrice Gerli, Ameth Hady Seydi, Michael Hamp, Nabil Mahaini, Norman Messer, Luyaku Nsimpasi, Elaine Reinke, Philippe Rémy, Momodu Sesay, Mohamed Shaker Hebara and David Suttie. Boris Gandon, Enrique Nieto from FAO and Nawsheen Hosenally from CTA contributed to the selection and write-up of the examples.

The following CTA colleagues provided inputs on CTA's work on youth and on the selection of case studies: Judith Francis, Jose Filipe Fonseca, Olu Ajayi, and Nawsheen Hosenally from the ARDYIS project (who contributed in addition to editing some stories). Clare Pedric (Consultant) drafted some initial articles and other inputs were shared by some CTA contacts including Anju Mangal, Miriama Kunawave Brown, Horace Fisher, Nomvula Dlamini, Diana Francis, Alex Njeru, Tukeni Obasi, Sithembile Ndema Mwamakamba and Gregg Rawlins.

The team benefited greatly from the information submitted by the respondents to the online survey and in particular by Maria Elenice Anastácio, Francis Xavier Asiimwe, Raynard Burnside, Aurélie Charrier, Magali Delomier, Franck Devort, Pavlos Georgiadis, Sophia Hellmayr, Ben Hillman, Wen-Chi Huang, Zernash Jamil, Robert Kibaya, Saing Koma Yang, Mariam Mansa Camara,

Michael McVerry, Milla Menga, Mireille Moko, Edward Mukiibi, Kakha Nadiradze, Sourou Hervé Appolinaire Nankpan, Anaclet Ndahimana, Moses Nganwani Tia, Marceline Ouedraogo, Romain Ouedraogo, Sopheap Pan, Pramesh Pokharel, Pascale Rouhier, Jean-Claude Sabushimike, Ameth Hady Seydi, Natalia Skupska, Tola Sunmonu, Sherman Tang, Kok Tha, Eloï Toi Tegba, Tong Chan Theang, Karen Tuason and Albert Yeboah Obeng.

The co-publishing agreement was arranged with the support of Bruce Frederick Murphy and Danila Ronchetti from IFAD; Lisa Pace, Marta Pardo and Rachel Tucker from FAO; and Jenessi Matturi from CTA. Final editing  was done by Ruth Duffy, and layout and design by Suzanne Redfern and Fabrizio Puzzilli.

# Executive summary

Global population is expected to increase to 9 billion by 2050, with youth (aged 15–24) accounting for about 14 percent of this total. While the world's youth cohort is expected to grow, employment and entrepreneurial opportunities for youth – particularly those living in developing countries' economically stagnant rural areas – remain limited, poorly remunerated and of poor quality. In recognition of the agricultural sector's potential to serve as a source of livelihood opportunities for rural youth, a joint MIJARC/FAO/IFAD project on Facilitating Access of Rural Youth to Agricultural Activities was carried out in 2011 to assess the challenges and opportunities with respect to increasing rural youth's participation in the sector. Over the course of the project, six principal challenges were identified. For each challenge, this publication presents a series of relevant case studies that serve as examples of how this challenge may be overcome.

The first principal challenge identified is youth's **insufficient access to knowledge, information and education** [Chapter 1]. Poor and inadequate education limits productivity and the acquisition of skills, while insufficient access to knowledge and information can hinder the development of entrepreneurial ventures. Particularly in developing countries, there is a distinct need to improve young rural women's access to education, and to incorporate agricultural skills into rural education more generally. Agricultural training and education must also be adapted to ensure that graduates' skills meet the needs of rural labour markets. Case studies from Cambodia, Uganda, Saint Lucia, Pakistan, Madagascar, Brazil, Ghana, Kenya, Rwanda and Zambia illustrate innovative ways of making this happen.

The second challenge identified during the project is youth's **limited access to land** [Chapter 2]. Although access to land is fundamental to starting a farm, it can often be difficult for young people to attain. Inheritance laws and customs in developing countries often make the transfer of land to young women problematic, and so are in need of amendment. Loans to assist youth in acquiring land are also needed, while leasing arrangements through which youth gain access – though not ownership – to land may also prove effective. Case studies from the Philippines, Burkina Faso, Ethiopia, Mexico, Egypt and Uganda all highlight possible means of improving youth's access to land.

**Inadequate access to financial services** [Chapter 3] was identified as the third principal challenge. Most financial service providers are reluctant to provide their services – including credit, savings and insurance – to rural youth due to their lack of collateral and financial literacy, among other reasons. Promoting financial products catered to youth, mentoring programmes and start-up funding opportunities can all help remedy this issue. Encouraging youth to group themselves into informal savings clubs can also prove useful in this respect. Case studies from France, Canada, Uganda, Moldova, Senegal, Cambodia, Bangladesh and Grenada all offer examples for policymakers and development practitioners of how rural youth's access to financial services can be improved.

**Difficulties accessing green jobs** [Chapter 4] was identified as the fourth challenge to strengthening youth's involvement in agriculture. Green jobs can provide more sustainable livelihoods in the long run, and can be more labour intensive and ultimately involve more value added. However, rural youth may not have the skills (or access to the necessary skills-upgrading opportunities) to partake in the green economy. Improving youth's access to education and training – including formal and informal on-the-job training – is needed to redress this skills mismatch. Case studies from the Zanzibar Archipelago, Rwanda, China, the United States, Bahamas, Kenya and Uganda all illustrate innovative ways of improving youth's access to the skills and opportunities needed to generate green jobs in agriculture.

The fifth principal challenge identified was young people's **limited access to markets** [Chapter 5], as without such access youth will not be able to engage in viable and sustainable agricultural ventures. Access to markets for youth is becoming even more difficult due to the growing international influence of supermarkets and the rigorous standards of their supply chains. Young rural women in developing countries face additional constraints in accessing markets, due in part to the fact that their freedom of movement is sometimes limited by cultural norms. Improving access to education, training and market information can all facilitate youth's access to markets, with niche markets offering particularly significant opportunities for young farmers. Facilitating their involvement in (youth) producers' groups can be similarly beneficial in this respect. Case studies from Kenya, Ghana, southern Europe, the United States, Tanzania, Colombia and Benin all offer examples of how youth's access to markets can be improved.

The sixth challenge identified was youth's **limited involvement in policy dialogue** [Chapter 6]. Too often young people's voices are not heard during the policy process, and so their complex and multifaceted needs are not met. Policies often fail to account for the heterogeneity of youth, and so do not provide them with effective support. To remedy this, youth need the requisite skills and capacities for collective action to ensure that their voices are heard. Policymakers themselves must also actively engage youth in the policymaking process. Case studies from Togo, Nepal and Brazil, as well as regional-level examples from Africa and Europe, all highlight ways to involve youth in shaping the policies that most affect them.

Addressing these six principal challenges will prove vital to increasing youth's involvement in the agricultural sector, and ultimately addressing the significant untapped potential of this sizeable and growing demographic. In developing countries in particular, facilitating the youth cohort's participation in agriculture has the potential to drive widespread rural poverty reduction among youths and adults alike. While these challenges are complex and interwoven, a number of key conclusions can be drawn from the case studies: ensuring that youth have access to the right information is crucial; integrated training approaches are required so that youth may respond to the needs of a more modern agricultural sector; modern information and communications technologies offer great potential; there is a distinct need to organize and bring youth together to improve their capacities for collective action; youth-specific projects and programmes can be effective in providing youth with the extra push needed to enter the agricultural sector; and a coherent and integrated response is needed from policymakers and development practitioners alike to ensure that the core challenges faced by youth are effectively addressed. Indeed, a coordinated response to increase youth's involvement in the agricultural sector is more important than ever, as a rising global population and decreasing agricultural productivity gains mean that youth must play a pivotal role in ensuring a food-secure future for themselves, and for future generations.

# Acronyms

| | |
|---|---|
| AASC | Annual Agro-eco-tourism Summer Camp (Bahamas) |
| ABED | Advanced Beekeeping Enterprise Development (Tibet) |
| ACE | Audio Conferencing for Extension |
| ACP | Africa, Caribbean and Pacific Group of States |
| AEDM | Agri-Enterprise Development and Management Programme (Cambodia) |
| AESA | Agro-ecosystem analysis |
| AGRA | Alliance for a Green Revolution in Africa |
| AIDS | Acquired immunodeficiency syndrome |
| AMRSP | Association of Major Religious Superiors of the Philippines |
| ANPE | Agence Nationale pour l'Emploi |
| ANPFa | All Nepal Peasants' Federation |
| ARD | Agriculture and rural development |
| ASSP | Agricultural Services Support Programme (Zanzibar) |
| ASY | Songtaab-Yalgre |
| AU | African Union |
| BOAD | West African Development Bank |
| CAADP | Comprehensive Africa Agriculture Development Programme |
| CAFY | Caribbean Agricultural Forum for Youth |
| CAI | Community Agricultural Information |
| CAN | Citizen Action Net for Social Development (Cambodia) |
| CAP | Common Agricultural Policy |
| CARP | Comprehensive Agrarian Reform Program (Philippines) |
| CBO | Community-based organization |
| CBU | Coffee business unit (Colombia) |
| CCI | Conscious Capitalism Institute |
| CEDAC | Cambodian Center for Study and Development in Agriculture |
| CEJA | European Council of Young Farmers |
| CGIAR | Consultative Group on International Agricultural Research |
| CMCSG | Community Management Course Saving Group (Cambodia) |
| CNJ | National Youth Council |
| COA | Taiwan Council of Agriculture |
| COM-VIDAs | Commissions for the Environment and Quality of Life (Brazil) |
| CONTAG | National Confederation of Agricultural workers (Brazil) |
| COONAFAT | Togolese national cooperative for the pineapple value chain |
| COP | Conference of the Parties |

| | |
|---|---|
| CSA | Community Supported Agriculture (China) |
| CSOs | Civil Society Organizations |
| CTA | Technical Centre for Agricultural and Rural Cooperation |
| CTOP | Togolese Coordination Committee of Farmers Organizations and Agricultural Producers |
| CUZA | Cooperative Union of Zanzibar |
| CYOP | Community Youth Outreach Program (Bahamas) |
| CYTS | CAN Youth Team Saving Group (Cambodia) |
| DA | Development agency |
| DAAD | German Academic Exchange Service |
| DISC | Developing Innovations in School Cultivation (Uganda) |
| ESP | Social Protection Division (FAO) |
| ESW | Gender, Equity and Rural Employment Division (FAO) |
| ETTI | Eastern Tibet Training Institute |
| EU | European Union |
| FAO | Food and Agriculture Organization of the United Nations |
| FARNPAN | Food, Agriculture and Natural Resources Policy Analysis Network |
| FFS | Farmer Field School |
| FHF | Friends Help Friends saving group (Cambodia) |
| FIBA | Farm In Bytes Application |
| FIRA | Investment Fund for the Future of Agriculture (Canada) |
| FNN | Farmer and Nature Net |
| FNPEEJ | Fonds National pour la Promotion de l'Entreprise et de l'Emploi des Jeunes (Benin) |
| FRAQ | Quebec Young Farmers' Federation |
| FSF | Friendship Saving Federation (Cambodia) |
| FSP | Financial service provider |
| FTJER | Young Rural Entrepreneur and Land Fund Programme (Mexico) |
| GCARD | Global Conference on Agricultural Research for Development |
| GFAR | Global Forum on Agricultural Research |
| GHG | Greenhouse gas |
| GKP | Global Knowledge Partnership |
| GRET | Group for Research and Exchange of Technology |
| GTET | Green Technology and Eco-Tourism (Tibet) |
| GYIN | Global Youth Innovation Network |
| HAFL | Bern University of Applied Sciences, School for Agricultural, Forest and Food Sciences |
| HIV | Human immunodeficiency virus |
| HOOPSS | Helping Out Our Primary and Secondary Schools Project (Saint Lucia) |
| ICT | Information and communications technology |

| | |
|---|---|
| IFAD | International Fund for Agricultural Development |
| IFPRI | International Food Policy Research Institute |
| IICA | International Institute for Cooperation on Agriculture |
| IIED | International Institute for Environment and Development |
| IPM | Integrated pest management |
| IPRC | Integrated Polytechnic Regional Centre (Rwanda) |
| ITC | International Trade Centre |
| JFFLS | Junior Farmer Field and Life School |
| JICA | Japan International Cooperation Agency |
| KDO | Kissan Dost Organization (Pakistan) |
| KFBG | Kadoorie Farm and Botanic Garden |
| KICK | Kisumu Innovation Centre Kenya |
| KWA | Kissan Welfare Association (Pakistan) |
| KWC | Khawateen Welfare Council (Pakistan) |
| LSG | Ladies Saving Group (Cambodia) |
| MEC | Microenterprise credit |
| MFI | MicrofFinance institute |
| MIJARC | International Movement of Catholic Agricultural and Rural Youth |
| MORDI | Mainstreaming of Rural Development Innovations |
| NAJK | Dutch Agricultural Youth Organisation |
| NAYA | National Association of Youth in Agriculture (Dominica) |
| NDBP | National Domestic Biogas Programme (Rwanda) |
| NFCP | National Federation of Coffee Producers (Colombia) |
| NFF | National Farmers Forum (Togo) |
| NGO | Non-governmental Organization |
| NRC | Climate, Energy and Tenure Division (FAO) |
| NYRC | Ndola Youth Resource Centre (Zambia) |
| OAS | Organization of American States |
| OPC | Office for Partnerships, Advocacy and Capacity Development (FAO) |
| PAEI | Programme d'Appui à l'Emploi Indépendant (Benin) |
| PAFPNET | Pacific Agricultural and Forestry Policy Network |
| PCD | Partnerships for Community Development (China) |
| PMD | Programme Management Department (IFAD) |
| PROMER II | Project for the Promotion of Rural Entrepreneurship II (Senegal) |
| PROPAC | Subregional Platform of Peasant Organizations of Central Africa |
| PTA | Policy and Technical Advisory Division (IFAD) |
| REJEPPAT | Network of young producers and agricultural professionals of Togo |
| REST | Relief Society of Tigray (Ethiopia) |
| RME | Rural microenterprise |

| | |
|---|---|
| ROPPA | Network of Peasant Farmers' and Agricultural Producers' Organisations of West Africa |
| RUFORUM | Regional Universities Forum for Capacity Building in Agriculture |
| RUL | Rivall Uganda Limited |
| SACCO | Savings and credit cooperative |
| SAFIR | Support Service to Rural Finance |
| SFYN | Slow Food Youth Network |
| SGL | Solidarity group lending |
| SHG | Self-help group |
| SLAFY | Saint Lucia Agriculture Forum for Youth |
| SME | Small and medium enterprise |
| SNV | The Netherlands Development Organisation |
| SOF | Saving for Our Future (Cambodia) |
| SPC | Pacific Community |
| SYFN | Savannah Young Farmers Network (Ghana) |
| TFM | Task Force Mapalad (Philippines) |
| TIC Americas | Talent and Innovation Competition of the Americas |
| TRF | Taiwan Rice Farmers Co. Ltd |
| TRN | Thai Rural Net |
| TVET | Technical and vocational education and training |
| UEMOA | Economic and Monetary Union of West Africa |
| UN | United Nations |
| UNEP | United Nations Environment Programme |
| UNESCO | United Nations Educational, Scientific and Cultural Organization |
| UN-HABITAT | United Nations Human Settlements Programme |
| UNICEF | United Nations Children's Fund |
| UNRWA | United Nations Relief and Works Agency for Palestine Refugees in the Near East |
| UPA | Quebec Union of Agricultural Producers |
| USAID | United States Agency for International Development |
| VC4A | Venture Capital for Africa |
| VSLAS | Village savings and loan associations |
| WB | World Bank |
| WDA | Workforce Development Authority |
| WFP | World Food Programme |
| WHC | Women's household credit |
| WHO | World Health Organization |
| WOS | Women Open School (Pakistan) |
| WWF | World Wide Fund for Nature |
| YABT | Young Americas Business Trust |
| YEDF | Youth Enterprise Development Fund (Kenya) |

| | |
|---|---|
| YEF | Youth Entrepreneurship Facility |
| YELP | Youth Entrepreneur Loan Project (Bangladesh) |
| YES | Youth Entrepreneurship and Sustainability |
| YFF | Youth For Future Saving Group (Cambodia) |
| YoBloCo | Youth in Agriculture Blog Competition |
| YPARD | Young Professionals' Platform for Agricultural Research for Development |
| YPF | Youth Peasants Federation |
| YSEEP | Youth Socio-Economic Empowerment Project (Republic of Moldova) |
| YSEI | Youth Social Entrepreneur Initiative |
| YSL | Youth start-up loan |
| YVP | Youth Volunteerism Program (Bahamas) |

# Introduction

Global population is projected to reach 9 billion by 2050. The number of young people (aged 15 to 24) is also expected to increase to 1.3 billion by 2050, accounting for almost 14 percent of the projected global population. Most will be born in developing countries in Africa and Asia, where more than half of the population still live in rural areas (UNDESA, 2011). Rural youth continue to face challenges related to unemployment, underemployment and poverty. Despite the agricultural sector's ample potential to provide income-generating opportunities for rural youth, challenges related specifically to youth participation in this sector – and, more importantly, options for overcoming them – are not extensively documented. Furthermore, statistics on rural youth are often lacking, as data are rarely disaggregated by important factors such as age, sex and geographical location.

A joint *Mouvement international de la jeunesse agricole et rurale catholique* (MIJARC/IFAD/ FAO) project on Facilitating Access of Rural Youth to Agricultural Activities was carried out in 2011 to address this information gap and to take a closer look at the challenges that rural youth face while engaging in agriculture. The agricultural sector is seen as crucial to addressing the disproportionately high levels of youth unemployment, underemployment and poverty. Not only is the sector of vital importance to rural economies worldwide – and particularly in developing countries – it also possesses significant untapped development and employment creation potential. To help seize this potential, the project engaged rural youth informants from all over the world to identify a core set of challenges that should be overcome for youth to create or seize decent work opportunities[1] in rural areas and to reverse the rural exodus. Within the framework of the project, steps were taken to explore examples (MIJARC/IFAD/FAO, 2012) showing how one or more of these core challenges were successfully addressed.

The findings of the joint project showed the need to further document the challenges that youth face while engaging in agriculture. The present publication was initiated in response to the request from rural youth representatives involved in the project to further document examples on youth engagement in the agricultural sector to identify potential solutions to overcome these challenges.

The publication is the result of a strategic partnership between the Food and Agriculture Organization of the United Nations (FAO),[2] the Technical Centre for Agricultural and Rural Cooperation (CTA) and the International Fund for Agricultural Development (IFAD). These three organizations joined forces to share their respective strengths, contacts and expertise during the development of the publication.

Throughout the years, the three organizations involved in this joint publication have shown strong support to youth development work in line with their overall mandates and strategies.

---

[1] According to ILO, decent work involves opportunities for work that is productive and which delivers a fair income; security in the workplace and social protection for families; better prospects for personal development and social integration; freedom for people to express their concerns, organize and participate in the decisions that affect their lives; and equality of opportunity and treatment for all women and men.

[2] Three FAO divisions were involved, namely the Climate, Energy and Tenure Division (NRC), the Social Protection Division (ESP) and the Office for Partnerships, Advocacy and Capacity Development (OPC).

**FAO**'s mandate is to raise levels of nutrition, improve agricultural productivity, improve the lives of rural populations and contribute to the growth of the world economy. Support to rural youth and young agricultural producers has been part of FAO's work for the last four decades in the form of strengthening and expanding young people's capacities, knowledge and skills (through education and training) and rural employment creation, while engaging them in major policy debates at global level. Strategic Objective Three of FAO's New Strategic Framework 2010–2019, "Reduce Rural Poverty", recognizes that rural youth should be treated as a priority group when it comes to accessing decent employment opportunities.

**IFAD** combats rural hunger and poverty in developing countries through low-interest loans and direct assistance. In its Strategic Framework (2011–2015), IFAD recognizes that rural youth account for a very large proportion of the population living in poverty, and that young people represent the main driver of rural economies of developing countries. As requested by young women and men representatives of farmers' organizations during a special session of the 2012 Farmers' Forum, IFAD is fully committed to intensifying efforts to ensure that the services and products offered by its projects actually reach out to and benefit young rural women and men, enabling them to increase their access to key social and productive assets and raise their voice to realize their full potential.

Funded by the European Union (EU), **CTA**'s mission is to advance food and nutritional security, increase prosperity and encourage sound natural resource management in ACP (African, Caribbean and Pacific) countries. It provides access to information and knowledge, facilitates policy dialogue and strengthens the capacity of agricultural and rural development institutions and communities. Youth is a priority cross-cutting issue in the CTA Strategic Plan for 2011–2015 and CTA has been supporting their engagement in agriculture for over two decades. With its new Youth Strategy, the organization intends to strengthen its support for youth in agriculture policy, promoting youth engagement in value chains, in information and communications technology (ICT) use for agriculture and rural development (ARD), and in agricultural science and tertiary education.

Although the United Nations (UN) defines youth as those persons between the ages of 15 and 24,[3] for the purpose of this publication the definition of youth might differ from one country to another depending on cultural and local patterns, and also according to the project/mechanism/ system described in the various case studies. In addition, emphasis is placed on the notion of entrance into agriculture rather than on a specific age category. It is also recognized that rural youth are a heterogeneous group (Bennell, 2010) and that the challenges faced by young men entering agriculture are multiplied for young women.

The overall aim of this publication is to provide development practitioners, including youth leaders, youth associations and producers' organizations, with insights into plausible solutions to overcome core challenges, providing examples from various countries. The publication team invited more than 1 400 potential contributors from all over the world, mostly members of agricultural producers' organizations, youth associations and non-governmental organizations (NGOs) to reply to an online survey (see Annex I). In parallel, a desk review was conducted and meetings were held with FAO, IFAD and CTA staff to identify suitable examples for inclusion in the publication. Out of all the entries received, a selection committee[4] identified a total of 47 case studies to be included in this publication. These case studies were selected because they represent innovate ways of overcoming challenges and have not yet been shared with a wider audience. The

---

[3] UNESCO. *What do you mean by youth?* (available at http://www.unesco.org/new/en/social-and-human-sciences/themes/youth/youth-definition/).
[4] The selection committee consisted of: Francesca Dalla Valle, Charlotte Goemans and Tamara Van 't Wout from FAO; Anne-Laure Roy from IFAD; and Nawsheen Hosenally and Ken Lohento from CTA.

intention is to inform and inspire the reader, offering options for overcoming a specific challenge in a specific context. Efforts have been made to feature countries and case studies that have not yet been exhaustively documented. While this is an innovative aspect of the publication, it means that the impact of certain examples has yet to be evaluated and that each example should be seen in its specific context and not from a one-size-fits-all perspective. While the publication team decided to document examples from low-, middle- and high-income countries, it is important to acknowledge that youth in developing and developed countries are likely to encounter different challenges and possess different capacities to address them. The broad scope of this study is not meant to facilitate comparisons across countries, but rather to illustrate that youth unemployment and underemployment is in fact a global problem, and that agricultural development can contribute to remedying this issue, irrespective of the country under consideration. This publication is an initial step towards documenting successful youth initiatives in agriculture. Further research is still needed to provide more in-depth evidence from additional successful cases.

The publication is structured according to the six challenges identified by youth themselves. These were taken to be the main challenges faced by youth seeking greater participation in the agricultural sector, and they are as follows: [1] access to knowledge, information and education; [2] access to land; [3] access to financial services; [4] access to green jobs; [5] access to markets; and [6] engagement in policy dialogue. While some challenges apply to smallholders in general, youth typically face more profound challenges. A rapid overview of the youth-specific aspects of the challenges is followed by country examples that may be used and adapted to overcome them. Each chapter ends with overall conclusions gathered from the various country examples from one or more chapters. The conclusions of each chapter include reference to additional examples which, while not fully developed as a case study, may still be interesting for the reader to further explore.

# 1. Access to knowledge, information and education

**MAIN AUTHOR: CHARLOTTE GOEMANS**

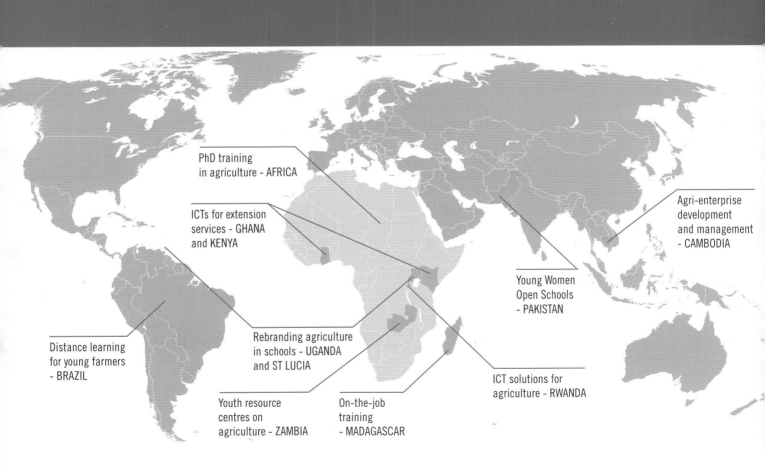

PhD training in agriculture - AFRICA

ICTs for extension services - GHANA and KENYA

Agri-enterprise development and management - CAMBODIA

Young Women Open Schools - PAKISTAN

Distance learning for young farmers - BRAZIL

Rebranding agriculture in schools - UGANDA and ST LUCIA

ICT solutions for agriculture - RWANDA

Youth resource centres on agriculture - ZAMBIA

On-the-job training - MADAGASCAR

# 1.1 Introduction

It is widely documented that education is key to overcoming development challenges in rural areas. Not only is there a direct link between food security and education of rural children, but it has also been shown that basic numeracy and literacy skills help to improve farmers' livelihoods (FAO, 2007). Youth's access to knowledge and information is crucial for addressing the main challenges they face in agriculture. In order for rural youth to shape agricultural policies affecting them directly, in terms of access to markets and finance as well as green jobs and land, they need to receive appropriate information and education. While this is true in developed and developing countries alike, it is of particular concern in the latter, where young rural inhabitants may lack access to even the most rudimentary formal education, and where educational institutions are often less developed. Formal primary and secondary education can provide young people with basic numeracy and literacy, managerial and business skills, and introduce youth to agriculture. Meanwhile, non-formal education (including vocational training and extension services) and tertiary agricultural education can offer youth more specific knowledge related to agriculture.

In developing countries, **access to information and education** is often worse in rural areas than in urban areas, and this discrepancy is observable as early as primary school. In many rural areas of developing countries, children are hungry and do not have the energy to attend school or easily absorb the information provided. During seasonal peaks in the agricultural cycle, there can be labour shortages and parents may see no other option than letting their children contribute to household and agricultural activities instead of attending school. The physical infrastructure of rural schools is often bad and classroom materials are sometimes lacking. Schools can be far away from rural communities making access difficult for rural children (FAO, 2009a). The United Nations Educational, Scientific and Cultural Organization (UNESCO) reports that rural children are twice as likely to be out of school as urban children (UNESCO, 2012). Although gender disparities in access to primary education have been greatly reduced, girls are still more likely to be excluded from primary education when they are poor and live in rural areas (UN, 2009). Ensuring the transition of rural children from primary to secondary school is an even bigger challenge for many developing countries, especially in sub-Saharan Africa.[5] Rural girls are less likely to attend secondary school than either rural boys and/or urban girls: early marriage limits girls' mobility and there tends to be a preference for the education of sons (UN, 2009; World Bank, 2011). Moreover, the curriculum is often not relevant to the rural context, and in many schools in developing countries, agricultural curricula have disappeared or are outdated and inadequate. In most parts of the world, agriculture is seen as a less worthwhile subject or as a last resort for underachievers, and using agricultural activities as a punishment is common practice in schools and households in many parts of Africa and the Pacific (MIJARC/IFAD/FAO, 2012; PAFPNet, 2010) – attitudes that negatively influence the aspirations of rural youth. The quality of education is often low, and good, motivated teachers willing to stay in remote rural areas are hard to find (FAO/UNESCO, 2003; World Bank, 2008).

In many rural areas, agricultural knowledge and farming know-how are passed on from parents to children. However, a survey carried out in the Pacific indicates that youth feel that such advice should be provided in a more coordinated and effective way, rather than on an informal basis (PAFPNet, 2010).

---

[5] In sub-Saharan Africa, only 64 percent of primary school students move up to secondary school (UNICEF Global Database, 2010).

**Vocational training and extension services** are potentially effective tools for teaching agricultural skills and providing capacity-building trainings for rural youth, but they do not always transmit the necessary skills, and so can result in poor employment outcomes (Bennell, 2007). Training programmes frequently lack funding and the capacities of service providers are rather weak. Low education levels among many rural youth further limit training possibilities (IFAD, 2010a). Furthermore, there is often a mismatch between the kind of training offered and the requirements of the labour market in an evolving agricultural sector (UNESCO, 2010). Rural youth repeatedly reported lack of training in areas such as leadership and business management as well as the need for apprenticeship opportunities.[6] In addition to these general constraints, training programmes mostly reach young men and do not cater to the needs of young women. This is particularly true in sub-Saharan Africa, the Arab states and in south and west Asia (Hartl, 2009). Root factors limiting young women's access to training include restricted mobility, young motherhood and limited schooling levels.

**Higher education** is equally essential for the development of the agricultural sector. The creation of high quality universities that focus on agricultural research and establish linkages with the farming community has proven beneficial for the development of the agricultural sector in countries such as Brazil, India, Malaysia and China (Blackie *et al.*, 2010). Connecting universities with farming communities is essential in order to broaden knowledge, increase research and development dissemination and enhance local problem-solving.[7] It is equally important to connect educational institutions with labour market opportunities and to build strong partnerships with employers to ensure that the skills of agricultural professionals respond to labour market needs so that young graduates are employable (Paisley, 2012). Unfortunately, in most developing countries, such systems are rarely instituted and access to tertiary agricultural education is low (FAO, 1997; Beintema and Di Marcantonio, 2009; World Bank, 2011a). In sub-Saharan Africa and Asia respectively, only 2 and 4 percent of university students are enrolled in agricultural studies (AFDB/OECD/UNDP/UNECA, 2012). Gender-disaggregated data on agricultural science and technology are scarce, but a recent IFPRI (International Food Policy Research Institute) study shows that women are underrepresented in agricultural research and higher education (Beintema and Di Marcantonio, 2010).

Modern **ICTs** such as mobile phones and the Internet are appealing to rural youth and have high potential for facilitating access to information to enhance productivity on the farm; enable agricultural innovation; and provide access to financial services and markets. Consultations with rural youth from all over the world made it clear that "youth pick up new technologies related to farming more easily and that young farmers are keen on increasing their production through improved and modern technologies" (MIJARC/IFAD/FAO, 2012). While mobile technology is generally widely diffused in rural areas, the Internet is not. High prices of computers and the Internet, combined with lack of electricity, limit access to the Internet in developing countries.[8] Rural women have less access to ICTs than rural men because of higher illiteracy levels and lack of financial resources to secure the use of ICTs (World Bank, 2011b).

---

[6]   Findings from the survey launched for the purpose of this publication (MIJARC/IFAD/FAO. 2012) and key informant interviews and findings from East African Farmers Federation Regional Youth Consultative Workshop, 2009 (Proctor and Lucchesi, 2012).

[7]   A WB evaluation of agricultural higher education programmes shows that the most successful projects linked the university with the community (available at http://lnweb90.worldbank.org/oed/oeddoclib.nsf/DocUNIDViewForJavaSearch/B6428A9B093BD7C6852567F5005D83CB).

[8]   Available data suggest that by the end of 2008 almost three-quarters of the world's rural inhabitants were covered by a mobile cellular signal, up from 40 percent in 2003 (ITU, 2010).

# 1.2 Case studies

## 1. AGRI-ENTERPRISE DEVELOPMENT AND MANAGEMENT

CAMBODIA

In order to support sustainable agriculture and rural development, the Cambodian Center for Study and Development in Agriculture (CEDAC) was set up in 1997 with the support of the French NGO, Group for Research and Exchange of Technology (GRET). In 2008 – with funding support from the Ecumenical Scholarships Programme, Hilfswerk der Evangelische Kirchen Schweiz, CODEGAZ and a Japanese private donor – CEDAC began implementing the Agri-Enterprise Development and Management (AEDM) programme in the provinces of Takeo, Kampot, Kampong, Chnang, Prey Veng, Svay Rieng, Siem Reap and Kampong Cham.

The AEDM programme aims to encourage out-of-school youth in rural areas to pursue professional careers in agriculture. The training programme lasts one year and covers the following areas: ecologically friendly agricultural techniques (e.g. mushroom and vegetable cultivation; chicken, pig and fish rearing); self-development and social education; farm management and business plan development; saving for self-reliance; financial management; and report writing skills. Month 1 is a probation period. Months 2 and 3 include basic training and an exposure trip to successful commercial farmers. Months 4 and 5, an internship is combined with field work and additional training sessions; internships are hosted by experienced farmers and entrepreneurs. Month 6 focuses on farm management and business plan development. During the subsequent five months, the business plan is implemented, accompanied by regular sharing/reflection seminars and further trainings. During the training programme, workshops with the trainee's parents are held to allow them to understand the details and benefits of the programme, in order that they might help and support their children.

The trainees are requested to pay back their training fee of USD 300 to CEDAC within three years of graduation. CEDAC uses this money as a revolving fund to provide loans to graduates lacking investment capital to set up their own farm business. Some trainees have established saving and credit groups to mobilize funds. The total saving capital of each group is around USD 2 000.

Since 2011, CEDAC has supported more than 300 rural youth aged between 16 and 30, and 40 percent are girls. Most have now started their own farm businesses or become community development workers or farm managers. On average, trainees' monthly income has increased from USD 100 to 300, reportedly two to three times higher than the income of youth who work in the clothing or construction sector. In addition, AEDM programme graduates have trained about 2 000 other youth in their communities.

(Case study drafted by C. Goemans, adapted from Survey)

Information provided by Mr Saing Koma Yang, President of CEDAC Cambodia.
More information available at:
www.cedac.org.kh
www.gret.org
www.ekd.de
www.heks.ch
www.codegaz.org

*(1) A young CEDAC trainee feeding pigs. (2) Mushroom growing. © CEDAC .*

## 2. REBRANDING AGRICULTURE IN SCHOOLS

UGANDA AND SAINT LUCIA

To encourage youth involvement in agriculture, two Ugandan graduates from Makerere University in Kampala, Edward Mukiibi and Roger Sserunjogi, founded the Developing Innovations in School Cultivation (DISC) project in 2006 with the support of Slow Food. The project aims to improve nutrition, environmental awareness and food traditions through the establishment of school gardens in ten primary and five secondary schools in Mukono. Students learn about food production from farm to table as they cultivate African indigenous vegetables such as amaranth, as well as fruits, in their school gardens and they learn how to cook from the school chefs. Students are taught to preserve seeds from local vegetable varieties so that they can use them for the next season and set up a seed bank. A "fruit and juice" party was organized for all the children of the participating schools so they could taste and learn about different fruits. The children's parents were impressed by their learning, reporting that their children are eager to go to school because of the fruits from the garden that they receive at lunchtime.

In Saint Lucia, the Helping Out Our Primary and Secondary Schools (HOOPSS) project was set up in 2009 by the Saint Lucia Agriculture Forum for Youth (SLAFY)[9] with the support of the Saint Lucia office of the Inter-American Institute for Cooperation on Agriculture (IICA).[10] The project has created school gardens in more than a dozen schools and teaches children techniques such as organic fertilizer use and rainwater harvesting. The HOOPSS initiative has a strong focus on marketing. Produce grown by schoolchildren goes to primary school feeding programmes, in an effort to tackle rising levels of hypertension, diabetes and obesity. Fruit and vegetables from the school farms are also sold to restaurants, hotels and the local supermarket chain, SUPER J, which has a commitment to buy regular supplies. SLAFY is urging for a percentage of the annual profits of the income from the school gardens to be shared among pupils – the money would be deposited

---

[9]    SLAFY is the Saint Lucia branch of the regional network the Caribbean Agricultural Forum for Youth (CAFY).

[10]   The HOOPSS programme was originally designed by the National Association of Youth in Agriculture (NAYA) of Dominica and is now also being implemented in Saint Kitts and Nevis.

in a savings account in the names of the individual students, to be accessed upon completion of their secondary schooling. In a recent initiative the school farm products were branded to increase national visibility and encourage parents and students to support the programme by purchasing HOOPSS items. A competition among students was held to design a logo for HOOPSS and by the end of 2013 it will be used on all products sold in supermarkets.

(Case study drafted by C. Goemans, adapted from CTA and Survey)

Information on project DISC provided by Mr Edward Mukiibi, Project Coordinator of DISC.
More information available at:
www.projectdiscnews.blogspot.com
www.slowfood.com/
https://www.facebook.com/pages/St-Lucia-Youth-Agricultural-Forum/274038609367030?ref=hl
www.iica.int

For other school garden projects, see also:
www.4hghana.org
https://www.facebook.com/pages/4-H-Ghana/338363636194265
http://4hghana.blogspot.nl/
www.jamaica4hclubs.com
http://www.fao.org/schoolgarden/index_en.htm

## 3. YOUNG WOMEN OPEN SCHOOLS

PAKISTAN

In Pakistan, cotton picking is done mostly by young women. They often take their small children with them when they work in the fields. Pesticides are increasingly sprayed in the cotton fields, sometimes right up to the harvesting period. Women and their children thus run severe health risks while in the fields. Compared with their male counterparts, women have less information and knowledge concerning the damage pesticides can cause to their health, while it is women and children who are most vulnerable to pesticides exposure.

In 2006, the World Wide Fund for Nature (WWF) launched a programme to address these issues. They established 42 Women Open Schools (WOS) in Southern Punjab and Northern Sindh provinces of Pakistan. The curriculum of the WOS was developed through consultative workshops with participants from different communities, including teachers and local government representatives as well as other NGOs active in the areas. In addition to pesticide risk reduction activities, health and hygiene and income generation activities were included in the curriculum. Vegetable seeds were handed out to women participants. Local farmers' organizations, including Kissan Welfare Association (KWA), Khawateen Welfare Council (KWC) and Kissan Dost Organization (KDO), also provided support for the implementation of the initiative.

In each WOS, around 20–25 women from the same or nearby villages come together to learn and to increase their confidence. Young literate women aged 18–22 years from farming families were trained to become Female Field Facilitators and run the WOS. These young women not only actively participated in the trainings but also convinced other women to join the WOS.

The WOS were highly successful in improving the safeguards used by the trained women. As a result of the training, young women cotton pickers now use protective gear and cover their body when they are in the sprayed fields, resulting in a 66 percent reduction in pesticide poisoning sickness. They stopped taking their children to the fields with them which led to increased well-being of the whole family. In the wake of these successes, the idea of WOS was further developed and Family Schools were established in over 100 villages in Southern Punjab and Northern Sindh. Both male and female community members attend these weekly Family School sessions, where topics related to decent work (e.g. freedom of association) are addressed. Other activities, such as kitchen gardening, environmentally friendly stoves and local embroidery, have been developed especially for young women members of these schools.

The training was successful because the courses were tailored to the needs of the local population and developed in a participatory way. Local young women were trained to lead the WOS, making the initiative more sustainable. Confidence-building activities empowered women to communicate to male family members the importance of attending public gatherings. The distribution of vegetable seeds was an additional motivation for families to let the women attend the training.

(Case study drafted by C. Goemans, adapted from Survey)

Information provided by Ms Zernash Jamil, Senior Programme Officer of WWF Pakistan. More information available at: www.wwfpak.org

## 4. ON-THE-JOB TRAINING

MADAGASCAR

The PROSPERER project began in 2009 with the support of IFAD. It promotes rural entrepreneurship for Malagasy youth through apprenticeships in the regions of Sofia, Itasy, Analamanga, Haute Matsiatra and Batovavy Fltovinagny. The project identifies rural microenterprises (RMEs) that could potentially host apprentices. Eligible RMEs must have premises located in the project regions, be operational, have the physical capacity to receive apprentices and already have experience with apprenticeships. RMEs can host between five and ten apprentices, depending on their capacity. The project provides training for the tutor or host RME to improve his/her training practices. Host RMEs receive USD 10 per apprentice per month for the purchase of materials needed for training.

Following identification of the RMEs, an information campaign is launched (through radio, word of mouth and household visits) to publicize the apprenticeships. Apprentices should be between 16 and 25 years old and priority is given to those from the most vulnerable families – a project validation committee ensures that the selection process is carried out correctly. Apprentices receive on-the-job training as well as separate training sessions on entrepreneurship and business management delivered by various external service providers. PROSPERER does not require any financial contribution from the young participants. Apprenticeships take place in subsectors, including silk weaving, basket weaving and beekeeping.

An apprenticeship lasts from two to six months, depending on the occupation. At the end of the practical training period, apprentices receive a certificate signed by the tutor and bearing the logos of the project and the chamber of commerce where applicable. Upon completion of the apprenticeship, graduates can choose between becoming a paid worker or an entrepreneur. For aspiring entrepreneurs, the programme provides start-up kits and support for young people creating their own business. To facilitate access to seed capital for graduates, PROSPERER has set up a guarantee fund with partner microfinance institutions (MFIs).

*(3 and 4) A weaving apprentice).*
*(5) Rearing of silk worms.* © IFAD

As a result of PROSPERER, 3 468 young people have started and 2 694 completed an apprenticeship. Some project trainees have been recruited as paid workers and others have become entrepreneurs. A project evaluation indicates that young women are more likely to turn to wage labour, while young men prefer to start their own business. Around 72 percent of the project beneficiaries were women, probably a result of the targeted sectors being mostly female dominated.

In 2012, PROSPERER launched a rural youth association strategy since it was found that young people who work in groups are better able to adapt to technological innovation allowing them to cultivate entrepreneurship. Also in 2012, the FAHITA network, a branch of the global network Global Youth Innovation Network (GYIN) was set up with the support of PROSPERER.

(Case study drafted by C. Goemans, adapted from IFAD)

Information provided by Mr Norman Messer, IFAD CPM for Madagascar.
More information available at:
http://operations.ifad.org/web/ifad/operations/country/project/tags/madagascar/1401/project_overview

Carton, M. & Rabezanahary, A. 2013. *Étude sur le terrain: La formation par apprentissage et l'inclusion des jeunes dans les activités non agricoles en milieu rural, dans le cadre d'une stratégie nationale de formation agricole et rurale. Réflexions sur la mise à l'échelle d'une expérience pilote à Madagascar. Experience Sheet*, Programme PROSPERER. *Promotion de l'emploi décent et productif des jeunes en milieu rural: Analyse des stratégies et programmes*. Sept. 2011. Programme annual reports.

## 5. PHD TRAINING IN AGRICULTURE

AFRICA

The Regional Universities Forum for Capacity Building in Agriculture (RUFORUM) is a consortium of 32 universities in eastern, central and southern Africa that was established in 2004. RUFORUM's mission is to enhance the quality and relevance of postgraduate education in Africa. It achieves this mainly through a competitive grants system which provides opportunities to postgraduate students and faculty members to be closely engaged in community action research. Three main types of grants are available,[11] and all proposals undergo rigorous external review, including review by the RUFORUM technical committee.

Since 2008, RUFORUM has implemented three regional masters programmes and six regional PhD programmes. The different doctoral programmes focus on Dryland Resource Management, Plant Breeding and Biotechnology, Aquaculture and Fisheries Science, Agricultural and Resource Economics, Soil and Water Management, and Agricultural and Rural Innovations. Each of these programmes is coordinated and hosted by one university – the "centre of leadership" – with other universities and knowledge centres intensely engaged in teaching, supervision, mentorship and curriculum development.

Students are trained in Africa, while being offered opportunity for international exposure. This means that five students can be trained in Africa for the cost of training one student abroad. Capacity is mobilized in the region for the running of the programmes, while external expertise and knowledge centres are involved through partnerships with, for example, the CGIAR (Consultative Group on International Agricultural Research) centres, ACP and EU knowledge centres and universities in Europe and America. All postgraduate programmes provide inbuilt support to students and staff for aspects often lacking in conventional curricula, such as skill enhancement courses (e.g. communication, IT, soft skills). For example, students with language difficulties can participate in language courses before starting their studies. An important component of the RUFORUM approach is the support to regional exchanges between staff and students, which helps to consolidate links and shared learning.

---

[11] Graduate research grants: Faculty staff members who are PhD holders can apply. The proposal/project must include training of at least two MSc students.
Community action research projects: Senior faculty staff members/professors can apply. The project must incorporate student training (at least one PhD and two MSc) and have a strong focus on community engagement.
Doctoral grants: Meant for PhD training, they include: (a) one-year finalization grants, mainly thesis completion, reporting back to the students' communities and dissemination through conferences, publications etc.; (b) two-year research grants for students starting their field research and registered in any of RUFORUM's regional programmes, with a preference for women and those registered in programmes outside their home country; and (c) full three-year scholarships, sponsoring students in any of the regional programmes.

The regional PhD programmes have received support for research, scholarships, short courses and staff exchange from a number of institutions, including the Bill and Melinda Gates Foundation, Rockefeller Foundation, International Development Research Centre, German Academic Exchange Service (DAAD), Alliance for a Green Revolution in Africa (AGRA), EU-EDULINK and EU-ACP Science and Technology Programmes, CTA and Carnegie Corporation.

A total of 100 students are currently enrolled in the PhD programmes. From the first group of 18 students who enrolled for the Dryland Resource Management programme in the last quarter of 2008, 15 have completed their studies. The most visible impact of the programme is the strengthened capacity – in research, teaching, mentorship, resource mobilization and networking – of both the individuals and the institutions/universities where they are employed.

(Case study drafted by C. Goemans, adapted from CTA)

More information available at:
www.ruforum.org

## 6. DISTANCE LEARNING FOR YOUNG FARMERS

BRAZIL

The National Confederation of Agricultural Workers (*Confederação Nacional dos Trabalhadores na Agricultura* – CONTAG), the largest labour union of rural workers of Brazil, initiated the Jovem saber (youth knowledge) programme in 2004 to enhance the skills of young farmers. In the framework of this programme, a free online training course has been developed for young farmers aged between 16 and 32. To participate in the training, the farmers have to form study groups of

*(6) Meeting of Jovem Saber participants from Brejo Santo.* © Ms Maria Elenice Anastácio

5–10 people, at least 30 percent of whom young women. The course has eight different modules and covers topics such as family farming, health, labour laws, and associations and cooperatives. Gender is a cross-cutting issue and the position of women in family farming and rural society in general is discussed throughout the course. The study groups have 45 days to go through each module on the Internet, after which they must hand in an assignment in order to move on to the next module. Additional face-to-face training and support activities are also offered by the CONTAG youth offices. To develop its activities, CONTAG receives financial support from the Ministry of Agriculture.

Since 2004, 26 000 youth have taken part in the training course, and CONTAG reports good success rates, with trainees obtaining access to land and agricultural credit, becoming union leaders and founding producers' organizations. Youth have been able to find local solutions to their problems and are engaging in policy debate to improve their situation in the long term. Furthermore, the training courses have given youth a stronger rural cultural identity and have improved their self-esteem.

(Case study drafted by C. Goemans, adapted from Survey)

Information provided by Ms Maria Elenice Anastácio, CONTAG youth officer.
More information available at:
www.contag.org.br

## 7. ICTs FOR EXTENSION SERVICES

GHANA AND KENYA

The Savannah Young Farmers Network (SYFN), a youth-led NGO, is running the Audio Conferencing for Extension (ACE) project in several communities in northern Ghana, offering innovative extension services. The ACE scheme addresses young farmers' special interest in approaching agriculture as a business, and helps them to explore some of the potential advantages – and pitfalls – of becoming agripreneurs.[12] Best of all, the system allows farmers to become involved in the content of the agricultural extension they receive. ACE is a two-way process, allowing farmers to ask questions about issues that interest them, and to steer sessions in the direction that will prove most fruitful for them.

With the help of a mobile phone, audio conferencing technology and a portable external loudspeaker, farmers in groups of 10–12 are linked up with agricultural extension workers and researchers, who between them offer a wide range of expertise. Although the experts may be some distance away, the technology enables farmers to seek their advice and ask questions about relevant issues. Trained community agricultural information (CAI) officers are on hand to ensure the smooth running of sessions, which generally take place twice a week within any given farming community – a huge improvement on the previous erratic services, when extension visits had to be made in person. An emergency meeting may also be called for, should the need arise.

Farmer feedback is key to this participatory process, and an important feature of the ACE approach is the use of videos filmed by CAI representatives, which document some of the challenges faced by farmers, and any solutions they might have developed. These short films are uploaded onto YouTube, or, in areas where there is no Internet, put on CD-ROMs. To date, videos have covered a wide range of subjects, including weeds, pests and diseases affecting crops and animals; model

---

[12]    "Agripreneurs" is a neologism for "agricultural entrepreneurs".

*(7) SYFN Executive Director Moses Nganwani Tia. (8) A farmer receiving agricultural advisory services via audio conferencing.* © Moses Nganwani Tia

farms; and post-harvest management. The documentaries have helped project staff develop specific extension advice to address the hurdles to greater productivity.

According to SYFN Executive Director, Moses Nganwani Tia, there is a direct link between the new extension advisory service and the number of young people taking up farming and starting their own rural enterprises. ACE not only provides technical information about crop growing and livestock keeping, but meets the strong demand for expert advice on marketing, credit and improved inputs and mechanization. Guests invited from both the private and public sector have made practical suggestions for further improving young farmers' access to the services and enabling them to succeed in agribusiness management.

By the close of the 2012 farming season, more than 880 farmers from 36 farmer groups across the three northern regions of Ghana had taken part in the audio conferencing sessions. The scheme had also reached a number of other value chain actors, including buyers, aggregators and representatives of financial institutions, as well as suppliers of farm machinery.

Another example on how ICTs (and tools such as YouTube) are used for extension services is the case of the MKulima Young online platform in Kenya, launched with the aim of encouraging youth to engage in agricultural issues, by connecting in a virtual space young farmers and youth aspiring to become farmers.

The platform members are mostly farmers and traders in agricultural produce and 95 percent are under the age of 32. Members can post questions on how to farm, receiving a prompt answer from other members; they can also buy or sell produce or any agricultural input online. The digital platform may also be used to receive information from experts or from those experienced in the farming of certain crops or livestock.

By using this digital hub and social media (Facebook, Twitter), youth are attracted to farms where they can earn a decent income. One young farmer who benefited from Mkulima Young membership explains: "When I have a question, produce to sell or something I need to buy, I post it there. Calls start to come in immediately."

(Case study drafted by C. Goemans and A. Giuliani, adapted from CTA and Survey, and from the Internet)

Information provided by Mr Moses Nganwani Tia, Executive Director of SYFN. More information available at:
http://ardyis.cta.int/en/news/other-news/item/144-an-attractive-opportunity
http://savannahyoungfarmers.wordpress.com/2012/06/20/savannah-young-farmers-network-syfn/
http://www.mkulimayoung.com/

## 8. ICT SOLUTIONS FOR AGRICULTURE

RWANDA

A high-tech lab in Rwanda is helping young people develop ICT-based agricultural solutions and turn them into commercial ventures. In June 2012, kLab was set up in Kigali to bring together tenants (the name given to young people with business ideas) and mentors (guest experts committed to helping the young innovators convert their dreams into reality). The link-up offers support to young entrepreneurs throughout the process, providing them with Internet access and training to improve their ICT skills and concrete advice on how to market their ideas and access venture capital. Workshops, camps and competitions help young people to become more commercial, teaching them how to write a business proposal, how to network and what kind of market to target.

Most of the young people who go to kLab are school or university leavers; others have already formed a start-up, but lack the business skills to market their product. In both cases, kLab is ready to help. Applicants are screened online and the ones with the most dynamic ideas are offered extensive support and guidance. So far, the lab has accepted about 70 members and helped shape some of their ideas into business ventures. Given that only 16 percent of kLab's tenants are women, kLab has initiatives designed to specifically target young women to join the tech business community such as the girls in ICT forum. The forum is chaired by 13 women, all tenants in kLab. To encourage fellow women to join the tech business, they organize public open talk events and hold debates in high schools and universities.

The business plans developed by teams of youth and mentors at kLab cover a range of sectors, including ICT applications for agriculture. A new and regular kLab feature invites farmers and heads of farmer cooperatives to talk about the challenges they face. Using this information, young people and their mentors will have more focus for the areas needing new ideas, and they will attempt to develop suitable ICT solutions in response.

Currently, mentors at kLab are working to help four high tech agricultural applications get off the ground. OSCA Connect, a start-up business group of young people, is developing a phone application called Sarura, which enables farmers to input the type of crop they wish to plant. The application then cross-checks meteorological data to determine if the crop is suitable, given the timing and location. Trials show that Sarura improves farmers' yields and saves them time and money. Another agricultural mobile and Web application undergoing development is Farm

*(9) A developer asking questions to farmers.* © kLab

In Bytes Application (FIBA). The application connects farmers to agronomists, business people and other stakeholders in the agricultural sector, and guides farmers in their day-to-day activities, helping them to update their farming records and profiles so that experts can offer more precise advice and targeted information. The third project is the creation of a Web site to provide farmers with information about crop cultivation and livestock resources, attracting advertising from the private sector and government organizations dealing with agriculture in the East African community. Finally, a Web application is being produced to link farmers' cooperatives with higher education students. The idea is for students to become attached to the cooperatives as volunteers or interns, providing the farmers with much needed technological skills and giving young people the opportunity to gain work experience.

(Case study drafted by C. Goemans, adapted from CTA)

More information available at:
http://klab.rw/

## 9. YOUTH RESOURCE CENTRES ON AGRICULTURE

The Ndola Youth Resource Centre (NYRC), a youth-led and youth-focused Zambian NGO, has set up and equipped seven youth resource centres that focus mainly on agriculture. The centres were established with grants from the Zambian Government at different times between 2000 and 2011 in Luapala Province, Northern Province, Muchinga Province and Copperbelt Province in Zambia. The centres have computers, printers and photocopiers, as well as telephones, television and radios. A small library offers a selection of books, newspapers and magazines.

Each centre is staffed by five instructors trained in ICT, business and agricultural development, who provide training, guidance, advice and information. Care is taken to research the labour market: on the basis of which agricultural skills and products are most needed, training programmes are drawn up and youths are offered appropriate support and guidance. Common themes are modern farming techniques and value chain processes. Strengthening business skills is an important part of the NYRC formula, with training sessions and individual coaching to help young farmers manage agriculture as a successful enterprise. Advice is given on aspects such as product selection, marketing and bookkeeping.

The centre links young farmers by e-mail, mobile phone text messages and radio to local partners such as the National Agricultural and Information service, the Zambia Agriculture Research Institute and the Organic Producers and Processors of Zambia. These linkages also enable young farmers to interact with each other and share experiences, challenges and solutions. Information management systems, installed at each youth resource centre, enable farmers to access information that will help them improve their production and marketing performance. The system links young farmers to a range of agricultural services that can provide information on weather, pests and pest control, seeds, inputs, soil testing and other issues. NYRC also works to improve youth's access to credit, providing assistance in business planning and proposal development, so that young farmers can apply for youth loans and credit under the government-run Youth Development Fund.

Currently, eleven youth farmer groups, each with around 45 members, use the services on a regular basis, attending their local youth resource centre where they receive continuous support. The young farmers of the Kwilimuna youth resource centre on the Copperbelt have found a good market for their agricultural produce, supplying products such as pork, vegetables, groundnuts and goat meat to leading super stores in the nearby city of Luanshya.

(Case study drafted by C. Goemans, adapted from CTA)

More information available at:
http://nyrcz.org/

# 1.3 Conclusions

For youth to successfully participate in the agricultural sector, access to both information and education are crucial. In addition to knowledge of agricultural production and processing techniques and the relative know-how, young farmers need access to information about finance, land and markets. This applies to developed and developing countries alike. However, the situation is particularly dire in many developing countries, where access to appropriate education and training often remains quite limited in rural areas.

In developing countries, young rural women tend to have particularly poor access to both general education and education that integrates agriculture sector knowledge. To meet this challenge, various strategies are implemented:

– introduction of a quota system, for example, gender quotas for training programmes in Brazil [case 6];

– direct targeting of women as participants, for example, training targeted specifically at women in Pakistan [case 3]; and

– provision of take-home food rations and introduction of flexible school calendars as incentives for families to let young women participate in training programmes.[13]

Access to tertiary education related to agriculture can be enhanced through scholarships, complemented by strengthening the capacities of universities in developing countries, for example, RUFORUM [case 5]. In order to ensure that the competencies of agricultural graduates meet the needs of an evolving agricultural sector, mechanisms can be introduced to facilitate close collaboration between educational institutions and local farming communities (Paisley, 2012).

Education in rural areas needs to be made more relevant by, for example, including agriculture in primary and secondary school curricula or modernizing the existing agricultural curricula. In Saint Lucia and Uganda [case 2], practical activities associated with the integration of agriculture in the school curriculum through school gardens successfully informed youth about the different ways of engaging in the agricultural sector, a potentially lucrative career choice. Showcasing the career paths of successful young farmers and "agripreneurs" as exemplary models can encourage youth to engage in the agricultural sector. The installation trophies initiative in France [case 17 discussed in Chapter 3], various CEJA showcasing initiatives [case 43 discussed in Chapter 6], as well as CTA's YoBloCo (Youth in Agriculture Blog Competition)[14] and the blog on '25 Youth that make agriculture "cool" by the Worldwatch Institute,[15] all help publicize and advocate youth engagement in agriculture. They may help reduce the stigma associated with agriculture as a no-win occupation full of drudgery, and instead highlight the broader possibilities of the agriculture sector as a source of thriving occupations.

---

[13]    For example, WFP (World Food Programme) operations in Somalia (http://www.wfp.org/countries/somalia/operations) and the opening of second chance schools in Morocco where holidays coincide with harvesting season (Maroc TelQuel Magazine Online, No. 316).

[14]    http://ardyis.cta.int/yobloco/

[15]    http://blogs.worldwatch.org/25-youth-making-agriculture-cool/

New training approaches for youth have been established, focusing not only on agriculture in the strictest sense, but on "sustainable socio-economic entrepreneurship", including the development of human skills (e.g. cultural, social, technical, organizational and economic) and the linking of agriculture to industry and services. The Songhaï Centre is widely known for this training approach,[16] which is also adopted by CEDAC, the Don Bosco Agro Mechanical Technology Center in the Philippines,[17] the Tutu Rural Training Centre in Taveuni Island of Fiji (McGregor *et al.*, 2011) and the Maisons Familiales Rurales in Europe, Africa, Asia and Latin America.[18] The FAO Junior Farmer Field and Life School (JFFLS) approach can also be cited in this context [case 27 discussed in Chapter 4].[19] Training needs to be tailored to the farmer and training curricula developed in a participatory way, as is the case in Pakistan, where youth from different communities took part in consultative workshops to define the activities of the Women Open Schools [case 3].

Modern ICTs have high appeal to young rural individuals. The Internet is becoming an increasingly important medium, including in the poorest regions of the world, to acquire access to information and allow producers to be better connected [case 33]. ICTs can also contribute to agricultural innovation, and the Rwanda case study [case 8] demonstrates that youth are potential pioneers in the development of ICT-based agricultural solutions. ICTs have also been instrumental in the development of new training techniques. In Brazil, a free online training programme for farmers is operational [case 6], and in Ghana, young farmers use ICTs for extension services [case 7]. In both cases, the use of ICTs is combined with face-to-face support in order to ensure that the tools are adapted to the needs of the target group. The experiences of the Grenada goat dairy project [case 24], Mfarm [case 35] and Vivuus Ltd [case 37] show how ICTs can facilitate access to financial services and markets. In various developing countries, rural information centres such as the Ndola Youth Resource Centre in Zambia [case 9] have been set up in order to improve access to modern ICTs.

Agricultural education and training must reflect the needs of the agricultural labour market and enhance the familiarity of young women and men with the world of work, including its practical challenges and rewards. In Cambodia, China and the Bahamas, internships and exposure trips are offered to trainees [case 1, case 31, case 26].[20] Young Malagasy are offered apprenticeships and support to set up their own businesses or become paid workers after completion of the programme [case 4]. In Rwanda, youth receive customized advice and are linked up with business partners and farmers so that their ICT-based agricultural solutions can fit the needs of the users [case 8].

The challenges related to access to information and education are complex. Ministries of education should work with a range of rural stakeholders including other ministries, the private sector, NGOs and producers' organizations to identify context-specific solutions. The capacity of service providers should be built and innovative integrated training methodologies should be explored. In order to improve the quality of education and to attract youth to the agricultural sector, youth should be directly linked to business partners and they should be given hands-on experience.

---

[16]  The Songhaï Centre started with sites in Benin and was replicated in other African countries, such as Nigeria, Liberia, Sierra Leone and Congo. The centre was nominated "Regional Center of Excellence" by the Economic Community of West African States (ECOWAS) in 2009 (see www.songhai.org).

[17]  http://www.dbagrolegazpi.org/

[18]  http://www.mfr.asso.fr/pages/accueil.aspx

[19]  http://www.fao.org/knowledge/goodpractices/bp-gender-equity-in-rural/bp-junior-farmer-field/en/

[20]  See also the FAO youth empowerment programme in Liberia, where FAO is training and offering internships to recent agriculture graduates of Liberian universities and the Songhaï Centre Liberia. Upon completion of the internship, interns receive letters of recommendation and are stimulated to develop a career in the agricultural sector (FAO Liberia Newsletter. *FAO Liberia's youth empowerment programme*, No. 5, July 2012).

# 2. Access to land

**MAIN AUTHOR: CHARLOTTE GOEMANS**

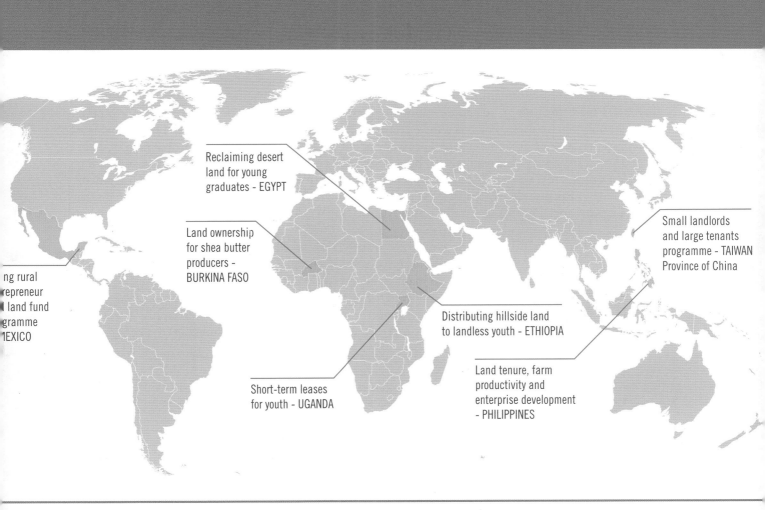

Reclaiming desert land for young graduates - EGYPT

Land ownership for shea butter producers - BURKINA FASO

ng rural repreneur land fund gramme 1EXICO

Short-term leases for youth - UGANDA

Distributing hillside land to landless youth - ETHIOPIA

Land tenure, farm productivity and enterprise development - PHILIPPINES

Small landlords and large tenants programme - TAIWAN Province of China

# 2.1 Introduction

Access to land is extremely important for young people trying to earn a livelihood in agriculture and rural areas. Land access is not only the number one requirement for starting farming, but it can also contribute to household food security and is a means for employment creation and income generation. Youth participating in the joint MIJARC/IFAD/FAO project reported that access to land serves as security and collateral for accessing credit, marks youth' identity, upgrades their status, and often enables participation in community decision-making organs and producers' organizations (MIJARC/IFAD/FAO, 2012).

Youth from all around the world see secure access to land as fundamental for entering farming,[21] yet they face greater challenges than adults. Moreover, the challenges faced by young men in accessing land are multiplied for young women. FAO reports that women constitute only a small proportion of all farmholders and that women typically hold smaller plots of land than men (FAO, 2011a). Although the challenges faced by young people are poorly documented and vary between regions and countries (and even within countries), it is possible to identify certain recurring issues.

Young men reported that the principal mechanism for accessing land is through inheritance (MIJARC/IFAD/FAO, 2012). Since life expectancy has increased in all regions, land transfer often happens at a later age and young men have to wait many years before inheriting their share of the family land, if at all. It is rare to encounter land transfer *inter-vivos* in developing countries, because land ownership is perceived as an adult privilege. On the one hand, youth are expected to wait until adulthood to own land. On the other hand, many young men are delaying marriage because they lack their own plot of land. However, in many parts of Latin America and the Pacific, tradition prescribes that one can only own land once one has established a family of one's own,[22] while in many parts of Africa, it is taboo for young people to access the family land while the parents are still alive (UN-HABITAT, 2011). While waiting for their inheritance, many youth just enjoy subsidiary land rights and work on the family land for little or no remuneration. In many developing countries, women do not inherit land and only obtain user rights via a male relative. Several countries have reformed their formal law system so that women are granted equal property and inheritance rights, but the enforcement of these formal laws can be very challenging, also because parallel customary law systems might exist denying equal land access for women. It can be especially difficult for young women to request enforcement of formal laws because they often lack the required knowledge, financial resources and confidence to protest against social norms and traditions (FAO, 2011a; World Bank/FAO/IFAD, 2009).

In high- and some middle-income countries, transferring farm land from one generation to the next usually means transferring the whole agricultural business on the land – a potentially complex process, including transmission of the management and ownership of the land, the business and other assets. Therefore, the parties involved often invoke family, financial, farm management and tax expertise to make sure that they avoid unnecessary transfer taxes and apply a transfer mechanism tailored to their specific situation (e.g. trust, family limited partnership).[23] The interests

---

[21] Findings from the survey launched for the purpose of this publication (MIJARC/IFAD/FAO, 2012) and key informant interviews and findings from East African Farmers Federation Regional Youth Consultative Workshop, 2009. (Proctor and Lucchesi, 2012).
[22] Findings from the survey launched for the purpose of this publication (Dirven, 2010).
[23] NJ State Agriculture Development Committee, Farm Link Program. *Transferring the family farm: what worked, what didn't for ten New Jersey Families.*

of two parties should be taken into consideration: those of the younger generation that need mostly financial support and training to take over the land/business, and those of the older generation that need financial security when handing over the farm assets.

Land management systems might change over time. Where land used to be owned by the community, lineage or clan, the control and management of land is becoming increasingly individualized. In developing countries, poverty often forces parents to sell their land to outsiders, excluding younger community members from land access. Large-scale land deals are particularly unfair towards young people, given that they are often not even consulted on agreements which may bar their and the next generations' access to land (White, 2012). In densely populated countries such as Rwanda, land has been highly fragmented and laws adopted prohibiting any further division of land. In practice, this means that the eldest son is the sole family heir and the final decision maker (IFAD, 2010b). What is more, increasing land degradation (FAO, 2011b) further limits the arable land available for young people.

It is unrealistic to expect youth to purchase land through acquired savings, given high rates of youth unemployment, low wages for most rural youth and high land prices. For young women in developing countries it is an even greater challenge to obtain the necessary capital to buy land as they often do unremunerated household work or subsist on low wages (FAO, 2011a). In addition, loans to buy land are not easily accessible for rural youth [chapter 3]. Land lease and rental are currently being explored to facilitate youth's access to land [case 15, case 16]. Furthermore, youth often lack knowledge on the existing land tenure systems in their area, which is not surprising as these systems can be a highly complex set of overlapping rules, laws, customs and traditions. Youth are not always aware of acquisition, registration and taxation measures, and so are disproportionately affected by corruption and the fraudulent activities of land dealers (UN-HABITAT, 2011).

Youth land rights are frequently not included in policy and legal documents and if they are included, no concrete implementation mechanisms are in place. Young people are not involved in the drafting of policies and laws related to land and find these frameworks unresponsive to their needs. In response to this challenge, FAO's Voluntary Guidelines on Land Tenure state that "effective participation of all members, men, women and youth, in decisions regarding their tenure systems should be promoted through their local or traditional institutions" (FAO, 2012).

# 2.2 Case studies

## 10. LAND TENURE, FARM PRODUCTIVITY AND ENTERPRISE DEVELOPMENT

PHILIPPINES

Despite the fact that the Philippine Constitution mandates agrarian reform with the aim of improving the land control of smallholders and rural workers, landownership is still in the hands of a few powerful and influential families (Llanto and Ballesteros, 2003). Given the inefficient implementation of agrarian reform and cases of violence and threats towards farmers who claim their land rights, there is a pressing need for farmers to improve their negotiating and campaigning capacities.

Task Force Mapalad (TFM), a national federation of farmers, farm workers and individuals working for agrarian reform and rural development, has been trying since 2001 to improve access to land

*(10) A young farmer grafting cacao in the Farmer Field School. (11) A young cacao farmer.* © TFM

through government programmes such as the Comprehensive Agrarian Reform Program (CARP) in five Philippine provinces.[24] In the past 11 years, TFM has facilitated the training of about 3 665 farmer paralegals and public communicators, in order that farmers may shield themselves from exploitation by various transnational agribusinesses, mining corporations and private individuals/big landowners. Paralegals strive to increase farmers' awareness of land rights through information and education campaigns, they promote potential alliances within and outside target communities, and engage in the formation and development of local organizations and partners. The activities of the paralegals benefit youth in particular since they often lack information about their land rights. Public communicators conduct advocacy campaigns and approach influential personalities in an attempt to place asset reform and rural development on the national agenda.

Both young men and young women take active part in the advocacy and training activities of TFM. In the Philippines, a farming initiative usually involves not only the parents but also their children. Young farmers in the five TFM target provinces are mostly seasonal farm workers because they

---

[24]    Provinces of Negros Occidental and Negros Oriental in the Visayas (central Philippines) and Bukidnon, Davao Oriental and Agusan del Sur in Mindanao (southern Philippines).

have few other work opportunities. Many are forced to move to the cities, where they are often engaged informally in the service sector, exploited as waiters, construction workers, drivers and house helpers. Improving land access for Philippine families means securing a future both for youth working on other people's farms and for youth moving away from agriculture.

TFM helps farmers not only to acquire land legally, but also to improve and manage lands once these have been transferred to them, and link them to markets. Youth are a special target of these farm productivity activities such as Farmer Field Schools (FFS) and farm technology trainings. In the past five years, TFM has produced 2 149 FFS graduates. TFM also provides support to youth engaged in enterprise development activities such as sugar-cane processing.

Alliance building and networking are key to TFM's activities. Partnerships have been established with religious institutions such as the Association of Major Religious Superiors of the Philippines (AMRSP) and with other civil society groups. TFM also has good relations with national and local government departments implementing the agrarian reform programme, including the Department of Agriculture and the Department of Agrarian Reform, as well as with different governmental and non-governmental agencies involved in agricultural enterprise and productivity development.

(Case study drafted by C. Goemans, adapted from Survey)

Information provided by Ms Karen Tuason, TFM focal point for women and youth.
More information available at:
www.taskforcemapalad.org
www.amrsp.org
www.da.gov.ph

## 11. LAND OWNERSHIP FOR SHEA BUTTER PRODUCERS

BURKINA FASO

Songtaab-Yalgre (ASY), which means "to help one another to a large extent", is a Burkinabe women's organization founded in 1990. ASY now counts over 3 000 members – 80 percent of whom young – from ten different provinces of Burkina Faso. The organization began as a small group of women who gathered to learn how to read and write, and gradually directed its efforts also to women's health issues and income-generating activities. ASY's members decided to pool their resources and to focus mainly on the harvesting and processing of shea nuts and the production and marketing of traditional and organic shea butter. Shea butter production is primarily a female activity, as traditionally young rural women are responsible for the collection of nuts fallen from the shea trees. ASY buys the shea nuts from its women members to transform them into shea butter.

In order to improve its members' economic empowerment, ASY started advocacy activities in the villages of Siglé, Boulsin and Gampela for women's access to land with a focus on young women. A group of women leaders is established, including someone close to the village chief. The group then meets with the village chief and with the land chief to negotiate land ownership so that the women members of ASY can grow shea trees. The negotiation process requires patience as it takes a minimum of two years and the importance of the planned activities and the ability of women to implement them must be demonstrated throughout. During the process, negotiations are held regularly and the women offer gifts to the elders and the village chief and also take part in the cultural and traditional activities of the village. As a result of ASY negotiations, more than 800 women (including 600 young women) are now the owners of a minimum 1-ha piece of land. The profits of the sale of ASY's shea products are distributed equally among its members. Women members of ASY who previously earned USD 1 per day as subsistence farmers now earn around USD 4 per day.

(Case study drafted by C. Goemans, adapted from Survey)

Information provided by Ms Marceline Ouedraogo, ASY Coordinator.
More information available at:
www.songtaaba.net
http://www.equatorinitiative.org/images/stories/winners/35/casestudy/case_1348150659.pdf

## 12. DISTRIBUTING HILLSIDE LAND TO LANDLESS YOUTH

ETHIOPIA

The Relief Society of Tigray (REST), an Ethiopian NGO, launched an initiative in 1999 to empower youth in the Hawezien District of the northern Tigray region of Ethiopia. More specifically, REST supported the construction of soil and water conservation structures on hillside lands. These structures were constructed by the whole community and then the rehabilitated land was distributed to landless youth in the Hawezien District. The young beneficiaries receive a landownership certificate from the village administration to make the land transfer official. In order to get youth started on their land, REST also supports other activities such as tree planting, beekeeping and the construction of water tanks in collaboration with local extension workers. The youth that benefit from the project are organized in associations which formulate their own land management by-laws in order to avoid misuse of the land.

Through the project, 360 landless youth have received a total of 90 ha of land and they now get an income through the sale of eucalyptus and honey produced on their land. In addition, they have fodder for animals and shrubs for firewood and fencing. Migration to urban areas has reportedly decreased since youth have their own income-generating activities in their communities.

(Case study drafted by C. Goemans, adapted from information provided by F. Dalla Valle)

More information available at:

http://www.rest-tigray.org.et/
www.drylands-group.org/noop/file.php?id=658

## 13. YOUNG RURAL ENTREPRENEUR AND LAND FUND PROGRAMME

MEXICO

Most of the land in Mexico used to be commonly owned. These common lands are called ejidos and are owned by *ejiditarios*. Tradition dictates that the transfer of rights on ejidos from one generation to another is heavily restricted; as a result, young households face land shortages, leading to youth migration to the United States. There is a widespread perception that older and less-educated ejiditarios with secure land rights are not very good at managing their land. To solve the problem, the Government decided to permit land sale on condition that the land be kept in the community.

To support youth taking over the ejidos, in 2004, the Government of Mexico together with the World Bank initiated the "Young Rural Entrepreneur and Land Fund Programme" (*Fondo de Tierras e Instalación del Joven Emprendedor Rural* – FTJER), with FAO as a project partner,

*(12) Tree planting.* © FAO

mostly providing technical assistance. The main objective of the programme is to support young farmers with entrepreneurial potential in acquiring the underutilized ejidos and other productive assets. The project also assists older landholders who transfer their lands to young farmers to access social welfare schemes for their retirement.

The Government of Mexico selected the poorest regions of the country as target areas for the programme. When the local authorities show interest in participating in the FTJER, an assembly is organized in the respective regions to disseminate information about the programme and to select youth participants. Workshops are organized for the eligible young entrepreneurs, where they learn about the details of the programme and the options for different project activities, identifying viable market opportunities in the local economy. Youth are then trained in the business area(s) identified and, together with a programme advisor, they develop business plans. The business plans are presented to Financiera Rural,[25] and if approval is granted, the Government provides loans through Financiera Rural to cover 100 percent of the price paid for the land. Loans are given at the interest rates applied by other commercial banks in the country and are designed to support only the most viable projects. FTJER staff provide legal advice for land transactions – for example, how land markets work and how to formalize contracts – as well as technical assistance for project implementation.

Between 2004 and 2008, more than 9 800 young people from 21 states of Mexico took part in the FTJER trainings, and about 4 000 received financial support from the programme to implement their projects. As a result, approximately 80 percent of the beneficiaries have obtained access to land, and of these, 70 percent through leasing transactions and 30 percent through purchase. About 90 percent of the land transactions took place among relatives, particularly between fathers and sons. The average beneficiary of the programme is a young head of household (approximately 31 years old) with a secondary education. Young women accounted for 43 percent of the total

---

[25]     Financiera Rural is a funding agency that provides credit to rural populations. Created by the Government in 2000, it replaced the Agricultural Development Bank (BANRURAL).

participants in the programme, and many of them reported that, thanks to the extra support from the programme, they managed to convince their husbands who had migrated to the United States to return to the village. Youth from indigenous communities also participated in the programme.

(Case study drafted by C. Goemans, adapted from the Internet and FAO)

More information available at:
Edouard, F. 2009. *Sistematización de experiencia juventud y tierra*. FAO.
WB. 2007. Returning Young Mexican Farmers to the Land. 2nd Annual Golden Plough Award for Innovative Project Design. In Agricultural and Rural Development Notes, issue 27, June.

## 14. RECLAIMING DESERT LAND FOR YOUNG GRADUATES

EGYPT

In the 1960s onwards, Egypt started providing free primary, secondary and higher education, resulting in increasing numbers of educated youth. In the 1980s however, the growing number of graduates had difficulty finding employment. Especially among graduates living in rural areas, unemployment rates were high.

In order to combat unemployment, boost the rural economy and have a secure source of food to feed the growing urban population, the Egyptian Government initiated schemes to grant bachelor degree graduates (giving priority to those with agricultural degrees) reclaimed desert land from the State. The beneficiaries were not only granted an irrigated plot of land but also a house with long-term credit and a certificate of ownership. The beneficiaries committed themselves to cultivate the land rather than sell it. The reclaimed state land is mostly located on the western side of the Nile Delta and receives water through irrigation canals from the Nile created and financed by the Government.

In 1993, IFAD in partnership with the Egyptian Government started the Newlands Agricultural Services Project providing agricultural support services (e.g. technology transfer, on-farm water management and credit) to about 35 550 small-scale farming households who settled in the reclaimed lands. As some graduates had no experience in agriculture, an extensive training programme, including demonstrations and excursions, was set up to address the difficult realities of farming and settling in the desert. In 2002, the IFAD West Noubaria Rural Development Project succeeded the Newlands Project and provided a loan to the Government of Egypt for approximately 71 400 ha of reclaimed land in Noubaria, with a 30-year repayment period and a grace period of five years. The graduates were then given the reclaimed land at a reasonable price to be paid back in comfortable instalments. Project activities included provision of food rations to new arrivals for four years and introduction of a drip irrigation system allowing farmers to diversify their crops and introduce new cash crops. The new irrigation method was successful, with the productivity of tomatoes increasing from about 4.5 tonnes/ha in 2003/04 to 7.5 tonnes/ha in 2006/07. A credit line was also created to support the development of micro- and small enterprises among graduates.

For the sale of produce, the project helped young farmers establish direct links with exporters and major buyers in the domestic market. They now supply fresh oranges and mozzarella cheese to resorts in Egypt's Sharm-el-Sheikh. Farmers also export sweet peppers and sun-dried tomatoes to Italy and the United States, peanuts to Germany and Switzerland, and raisins, artichokes, apricots, peaches and potatoes to a variety of European countries. Perhaps their most impressive contract is with Heinz, a global food company, which buys more than 6 000 tonnes of tomatoes each year from 300 project farms. Heinz provides the farmers with seeds of the required quality and guarantees the purchase of half the harvest at an agreed price. If the farmers cannot sell the remaining tomatoes in

*(13) A beneficiary of the IFAD West Noubaria Rural
Development Project with ready-to-sell oranges .
(14) A beneficiary of theIFAD West Noubaria Rural
Development Project with peanuts grown on reclaimed desert land.
(15) Drip irrigation system.* © Nabil Mahaini

the domestic market, Heinz is committed to buying them. In 2012, the project managed to export the produce of almost 3 000 ha consisting mainly of potatoes, tomatoes, peanuts and other fruits and vegetables.

The IFAD projects contributed to the development of young rural communities and supported the construction of health centres and schools. Desert land became more attractive to youth, services and infrastructures improved, and the sense of community was enhanced. As a result, between 2002 and 2012, the proportion of successfully settled graduates soared from 25 to 98 percent.

(Case study drafted by C. Goemans, adapted from IFAD)

Information provided by Mr Abdelhamid Abdouli (Former IFAD CPM Egypt), Mr Mohamed
Shaker Hebara (IFAD CPM Egypt) and Ms Elaine Reinke (IFAD Communication Officer).
More information available at:
http://www.ruralpovertyportal.org/country/voice/tags/egypt/egypt_desert
http://operations.ifad.org/web/ifad/operations/country/project/tags/egypt/306/project_overview
http://operations.ifad.org/web/ifad/operations/country/project/tags/egypt/1204/project_overview
http://www.youtube.com/watch?v=dW8_L8S2w8A
Adriansen, H.K. 2009. *Land reclamation in Egypt: A study of life in the new lands*. In Geoforum,
No. 40, Elsevier.

## 15. SMALL LANDLORDS AND LARGE TENANTS PROGRAMME

TAIWAN PROVINCE OF CHINA

The Taiwan Council of Agriculture (COA) launched the "Small Landlords, Large Tenants Programme" in 2008. The main purpose of this programme is to encourage elderly farmers to lease their land on a long-term basis to young farmers and to farmers' organizations. A farmland database has been set up by COA and provides information on and is a platform for the sale and lease of farmland so that interested buyers and sellers quickly find what they are looking for. The database is managed at local level by farmers' organizations which have proved excellent implementing partners. The elderly farmers who decide to lease their land are properly guided and included in a retirement system which allows them to provide consultancy to their tenants. In 2010, COA began training the young lessees in marketing and management and helped them improve their farm equipment and facilities through low interest loans. Youth can thus increase farm size and reduce production costs. Furthermore, youth interested in engaging in agriculture were offered internships in the programmes successful farms.

By the end of 2010, the programme had matched 8 121 landlords with 703 tenants, producing a rejuvenating effect. As a result of the programme, land tenants have on average 8 ha of land each, which is about seven times the land held by the average farm household in Taiwan Province of China. The programme also had an important effect on gender relations, because women could easily participate in the project as land tenants, while they would normally have difficulty claiming their land inheritance rights.

The Taiwan Rice Farmers Co. Ltd (TRF) is a successful example of the programme. TRF comprises 43 young farmers from seven different counties who have benefited from the Small Landlords, Large Tenants Programme and who farm leased land. They focus on a variety of high quality produce and engage in contract farming with individuals and businesses, developing an effective marketing strategy for their produce.

(Case study drafted by C. Goemans, adapted from Survey)

Information provided by Ms Wen-Chi Huang, Associate Professor of National Pingtung University of Science and Technology. More information available at: www.coa.gov.tw

## 16. SHORT-TERM LAND LEASES FOR YOUTH

UGANDA

Rivall Uganda Limited (RUL) is a trading firm, dealing in an extensive range of food grains, vegetable oils and honey. Given that most farming in Uganda is smallholder in nature, it is difficult and expensive for RUL to collect sizeable quantities of a particular commodity from one or a small number of locations. In order to overcome this constraint and to be able to fulfil supply contracts with regard to quantity specifications, RUL started working with youth groups from the Kisoro district in southwest Uganda in 2011.

RUL enters into short-term lease agreements with landowners that do not wish or do not have the capacity to utilize their land in the foreseeable 12 months. RUL works closely with the local authorities to sensitize landowners with regard to youth and their need for access to land. They stress that no transfer of land title is involved and that the lease will benefit the whole community. Landowners' children are encouraged to form or join groups to use their parents' or relatives' land. Once RUL and the landowner agree on the terms of the lease between RUL and the owner and on the method of payment (cash or a proportion of the produce), and once the Uganda Land Commission approves the lease, RUL communicates the availability of the land to current or prospective youth groups through noticeboards in the community and the RUL extension workers. Youth groups should at least have eight members (aged 18–35), three of which female, in order to be able to apply for the lease. Some preselected groups are then invited for an interview and may be allocated a plot of land. RUL also leases agricultural machines to the youth groups for tilling the arable land available to them.

RUL's extension officers, trained by National Agricultural Advisory Services, work closely with the youth groups from the sowing stage through to harvest and sale. The extension workers are all under the age of 30 – a factor which has been useful for understanding the challenges faced by youth and their mind-sets. RUL connects the groups to buyers and recovers payment from the sale of the produce. Working with groups rather than with individuals has been key to the success of the initiative. Aggregating youth in groups boosts morale and means that when some group members are unable to participate in farming the land, others will continue the work.

The mechanism is a win-win arrangement for all involved: the youth obtain a good income by pricing produce competitively; landowners receive cash/produce from their land that would otherwise be unused; and RUL obtains reliable supplies for its partners. In the past year, RUL has worked with 31 youth groups comprising 411 youths. They have established long-term supply arrangements with large entities that use agricultural produce, including hotels, supermarkets, schools, beverage companies and exporters.

(Case study drafted by C. Goemans, adapted from Survey)

Information provided by Mr Francis Xavier Asiimwe, Coordinator of Agricultural Extension and Research, RUL.
More information available at:
www.rivalluganda.com
http://www.naads.or.ug/

# 2.3 Conclusions

In order to facilitate youth's access to land, action is required on various fronts. The actions to be taken will depend on the prevailing issues in a given country, and so will vary both between developed and developing countries, and among developing countries. The case studies illustrate various mechanisms that respond to the needs of young men and women:

> **Advocacy** towards implementing existing laws and regulations granting youth access to land. In the Philippines [case 10], advocacy activities resulted in more efficient implementation of the agrarian reform mandated by the Philippine Constitution. In Burkina Faso [case 11], advocacy campaigns targeted the local traditional authorities, such as the village chief and the land chief, to release some of the land to young women.

> **Rehabilitation** and subsequent distribution of land to young people. This mechanism is especially suitable when there is a scarcity of arable land, as illustrated in the examples from Ethiopia and Egypt [case 12, case 14].

> **Provision of loans** specifically targeted at youth for acquiring land. In order to make sure that youth are able to pay back the loan, they should be properly guided and trained [chapter 3]. This holds true for both developed and developing countries, as limited access to credit for young entrepreneurs is a relatively universal problem. For example, in Mexico, youth received advice and training to draft viable business plans before they gained access to loans to acquire land for the implementation of their projects [case 13]. A similar approach has been adopted in the installation procedure in France [case 17].

> **Leasing** to provide youth with access to land. In Taiwan Province of China and Uganda [case 15, case 16], intermediaries (COA and RUL, respectively) encouraged landowners to lease their land to youth. Farmer and Nature Net (FNN), a farmers' organization from Cambodia, obtained long-term land lease contracts from landlords and then subcontracted the lease to young farmers' cooperatives. While FNN bought a rice mill to process the produce and financially supported the young farmers, the young farmers themselves invested their labour in the land. With the income from the sale of their produce, the young farmers are gradually taking over the lease from FNN.[26]

Most of the mechanisms listed above involve an intermediary – e.g. a women leaders' group in Burkina Faso or a private company in Uganda – to facilitate communication between the youth and the elderly regarding the transfer of land. As land issues are often extremely complicated and require changes in mentality, tradition and relations between young and old in the community, it is essential to involve all community members in the discussion process and to break the silos between generations. Where government resources allow, one option is to give older community members incentives to transfer (part of) their land to the young generations, as happens in Taiwan Province of China [case 15] and Mexico [case 13], where access to retirement systems is offered in exchange for land transfer. The EU Rural Development Policy 2007–2013 addresses both sides of the coin and proposes two measures that Member States can include in their rural development programme to facilitate the intergenerational transfer of farms. The first aims to financially support the setting

---

[26]    Information provided by Mr Sopheap Pan, Executive Director of Farmer and Nature Net (FNN) Cambodia.

up of young farmers under the age of 40, while the second encourages early retirement of farmers over the age of 55.[27] By 2010, 17 000 European farmers and farm workers had benefited from the early retirement schemes, releasing roughly 22 000 ha of farm land, and about 36 000 young farmers had received support to start up their own farms.[28]

In all types of land transaction, it is important to strengthen youth's awareness of land tenure systems and the relative judicial aspects. Some examples focus specifically on this, with farmers trained as paralegals in the Philippines [case 10] and project staff giving legal advice to youth on land transactions in Mexico [case 13]. In Ethiopia, the project provided youth with landownership certificates to make the land transfers official [case 12]. Rwanda has also adopted an innovative approach in this regard: conflict resolution committees (Abunzi) have been established and their members have been trained by the Ministry of Justice. Around 60 percent of the cases that the Abunzi deal with are land cases, 50 percent of which involve women. A minimum of 30 percent of the members of each Abunzi must be women. The Abunzi report that having a number of women on the committee lowers the threshold for women, and especially young women, to file their land cases (IFAD, 2010b).

Although the above-mentioned examples focus on access to land, they almost all also provide services (e.g. training) and inputs to raise the productivity of the land and improve the processing and marketing of agricultural produce. In this way, youth are supported not only in accessing land but also in making the land more productive and generating income.

**Rural institutions** play an important role in all the examples. In the Philippines, Task Force Mapalad has been carrying out its activities through its members, which are mostly farmers' organizations [case 10]. In Taiwan, farmers' organizations formed excellent managers of the Farmland Database [case 15]. Groups of women leaders have been instrumental in convincing the village chiefs to release land to young women in Burkina Faso [case 11]. In Uganda [case 16] and Ethiopia [case 12], youth have been encouraged to form or join organizations to manage their land. Local authorities are essential project partners, for example in Uganda, where they assist in convincing landlords to lease their land to youths [case 16].

---

[27]  Council Regulation (EC) No. 1698/2005 of 20 Sept. 2005 on support for rural development by the European Agricultural Fund for Rural Development (EAFRD).

[28]  European Network for Rural Development. Rural Development Programmes 2007–2013: Progress snapshot 2007–2010.

# 3.  Access to financial services

**MAIN AUTHORS: MARTINA GRAF AND FRANCESCA DALLA VALLE**

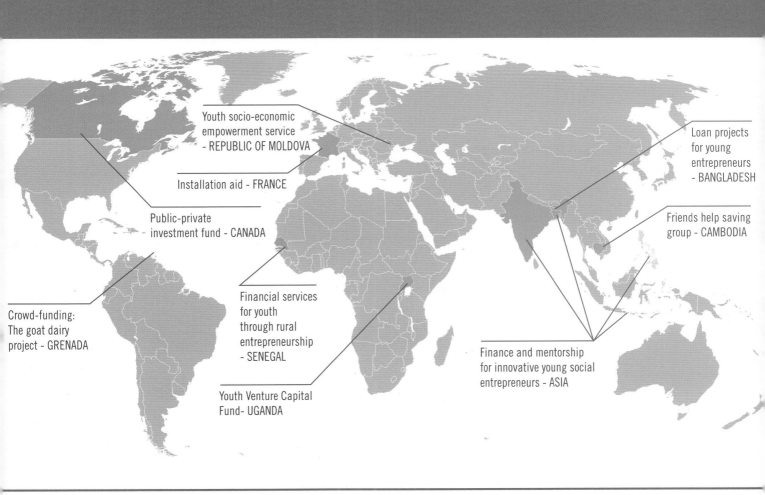

Youth socio-economic empowerment service - REPUBLIC OF MOLDOVA

Installation aid - FRANCE

Public-private investment fund - CANADA

Crowd-funding: The goat dairy project - GRENADA

Financial services for youth through rural entrepreneurship - SENEGAL

Youth Venture Capital Fund- UGANDA

Loan projects for young entrepreneurs - BANGLADESH

Friends help saving group - CAMBODIA

Finance and mentorship for innovative young social entrepreneurs - ASIA

# 3.1 Introduction

Just like access to land, access to financial services such as savings and loans is of fundamental importance to start any agricultural activity [chapter 2]. Even if youth do have access to land, they still need finance to cover the costs of planting and harvesting, as well as investments in improved productive capacities. Payment and trading services, such as mobile banking and Internet trading, are important tools for selling their produce. Moreover, given that the agricultural sector is often exposed to adverse natural events that negatively affect production (Dalla Valle, 2012), access to insurance schemes is crucial for young farmers.

In order to meet these needs, financial service providers (FSPs) (IFAD, 2010c) have to play a crucial role. FSPs include formal banking systems (commercial and development banks), semi-formal banking systems (savings and credit cooperative organizations [SACCOs]) and informal banking systems not officially registered at national level (e.g. self-help groups [SHGs], village savings and loan associations [VSLAs], moneylenders and traders). However, data from several leading FSPs reveal that young people tend to make up a smaller proportion of overall formal FSP clientele than their overall population demographics would suggest. Providing financial services in rural areas is typically considered high risk due to the unique characteristics of agriculture: dependence on natural resources and seasonality; long production cycles; and vulnerability to variable weather. Furthermore, scattered rural populations greatly increase the operating costs of financial institutions.[29]

While financial services have become increasingly available to poor farmers, there is still much to be achieved to improve the availability of such services to young people in agricultural and rural enterprises (Dalla Valle, 2012). In both developed and developing countries, most FSPs provide few savings or insurance services for youth, focusing more on credit, despite the fact that savings remain extremely important to youth for building up assets for investments and insurance (MIJARC/IFAD/FAO, 2012). In many countries, laws and regulations also exclude people below the age of 18 from accessing any financial products and services (UNCDF, 2012). In addition, few, if any, financial products are specifically tailored to youth (MIJARC/IFAD/FAO, 2012). Research shows that while the majority of micro-finance institutions (MFIs) do serve young people above the age of 18, they are rarely recognized as a specific client group, and few products are developed to meet their unique needs (Shrader *et al.*, 2006). Furthermore, an ongoing global debate demands the revision of MFI concepts, since many MFIs providing loans to youth often charge too high interest rates (UNCDF, 2012). Before releasing funds to youth, many FSPs ask for loan guarantees, such as formal land titles [chapter 2], steady employment, personal guarantors, solidarity group guarantees or more informal guarantees (motorcycles, furniture etc.) – all assets that youth typically do not possess. Furthermore, youth are perceived as a high-risk category (Atkinson and Messy, 2012) because of their limited financial capabilities,[30] often resulting from their lack of experience. Despite growing recognition of the importance of inclusive finance, there are few innovative models on the reduction of risk when lending to youth.

---

[29] Dalla Valle (2012); http://www.ruralfinance.org/fileadmin/templates/rflc/documents/8_Rural_finance_challenges_web.pdf.

[30] Financial capability: i) day-to-day management of finances, for example effective budgeting and use of a bank account; ii) planning ahead for retirement, other life transitions and unexpected events for example by saving; iii) knowing where, and how, to seek appropriate financial advice; and iv) financial literacy in general (Atkinson and Messy, 2012).

Therefore, youth often rely on informal sources – typically family and friends – to obtain access to financial services. Information about informal services in rural areas is mostly provided through informal channels (e.g. word of mouth and the radio) (Dalla Valle, 2012). Agro-processing companies, input suppliers and traders often supply credit for inputs or farmer insurance [chapter 5], but, like many MFIs, they often charge very high interest rates.

ICT offers a wide range of products for accessing financial services, such as e-banking, e-business and e-trade through mobile banking. Nowadays, a third of the world's population uses the Internet, and more than 45 percent of them are young people, mainly in urban areas. Many rural areas still lack Internet connections – particularly in developing countries – which can act as a drawback for young people wishing to stay in rural areas and conduct their business from there (Dalla Valle, 2012).

Another barely affordable financial service is agricultural insurance. If it were more affordable, it could support young farmers in developing better agricultural risk management strategies for their farms. To improve such services, appropriate policies should be drafted and existing services revised to reach a younger clientele (Dalla Valle, 2012).

Collective action is somewhat lacking among rural youth [chapter 5], and they are rarely organized in self-help groups that could provide the means for generating savings and improving the borrowing power of both individual members and the group. In many developing countries, young rural women face additional constraints in accessing financial services due to their higher rates of illiteracy, restricted liberty of action and lack of consent of family members, much of which can be traced to gender discrimination embedded in societal norms [chapter 1, chapter 2] (Dalla Valle, 2012).

# 3.2 Case studies

## 17. INSTALLATION AID

FRANCE

In order to ensure that the younger generation enters farming, the French Government, in collaboration with the EU, provides support to young farmers wishing to set up in business. The mechanism of installation aid has been in place since 1973. Since its creation in 1957, the French Young Farmers Syndicate (*Jeunes Agriculteurs*) has played a big role in the development of the installation procedure, supporting the cause of young farmers and lobbying for and promoting the installation procedure.

The installation procedure offers a range of options, and youth wishing to set up in farming can apply for three types of financial support: (1) an installation grant; (2) special loans at reduced interest rates; and (3) tax and other types of benefits. In order to benefit from the installation grants and special loans, the applicant must be under the age of 40 and adhere to specific conditions related to education. The applicant should also draft a customized professionalization plan and a business plan covering a five-year period. Advisors from the information points set up in the various departments assist the applicants with the drafting of these documents. The application is then assessed by various institutions such as the Departmental Commission for Agricultural Counselling. Most youth applying for installation aid wish to take over an existing agricultural business.

*(16) An installation trophy winner.* © Pixel Image / Laurent Theeten

In addition to the installation procedure, a specific tax law in favour of young farmers was approved in 2010 following the advocacy activities of *Jeunes Agriculteurs*. The law states that a specific tax is applied when agricultural land is sold for non-agricultural purposes. The money obtained from the application of this tax goes to a fund dedicated to the installation of young farmers.

In 2011, *Jeunes Agriculteur*s launched the Installation Trophies Project in collaboration with the National Federation of Fruit Producers. Young fruit producers who have been active for at least five years can participate by submitting a description of the activities carried out since installation. The career descriptions of ten young fruit producers are selected and put on the trophies Web site; three of them win a trophy. In 2012, the competition was extended to wine makers and market gardeners. The aim of the competition is to showcase the career paths of young people engaging in agriculture and to stimulate others to choose a similar profession.

(Case study drafted by C. Goemans, adapted from Survey)

Information provided by Ms Aurélie Charrier, Installation Advisor of Jeunes Agriculteurs.
More information available at:
www.jeunes-agriculteurs.fr
www.trophees-installation.com
www.gouvernement.fr
http://europa.eu/
Valentin, J., Bigand, J. & Guillaume, S. 2012. Une installation réussie pour tous.
46th National Congress of Jeunes Agriculteurs, Pontarlier.

## 18. PUBLIC-PRIVATE INVESTMENT FUND

The Canadian Government decided to give the younger generation access to resources to start their agricultural activities. In spring 2011, a USD 75 million public-private Investment Fund for the Future of Agriculture (*Fonds d'investissement pour la relève agricole* – FIRA) was launched for young Quebec farmers as a result of four years of lobbying by the Quebec Young Farmers' Federation (Fédération de la relève agricole du Québec – FRAQ) and the Quebec Union of Agricultural Producers (UPA). FIRA features three equal partners: the Quebec Government, the Desjardins financial institution and the pension fund of the Quebec Labor Union. Quebec's farm credit agency, *Financière agricole du Québec*, administers the fund. FIRA's mission is to support young people starting projects or agriculture businesses in all regions of Quebec through investments in the form of loans or agricultural land lease agreements.

The new fund is available to farmers aged between 18 and 40. Not only individuals, but also legal societies, such as collectives and cooperatives, can benefit from the financial services offered by FIRA and which take three different forms, as FIRA can: offer loans with flexible repayment terms,[31] take an equity stake in farms[32] or purchase land for leasing purposes.[33] A young farmer contracting with FIRA may also acquire shares or assets for the farm business, such as equipment, buildings, agricultural land, quotas or even a complete farm. If a loan is received, the repayment has to be initiated after three years, and a progressive and beneficial interest rate is applied.

The contractors must present a strong business and management plan, at least for the first five years of the operation at the start-up or transfer of an agricultural enterprise. After three years of activity, the young entrepreneur must demonstrate a minimum turnover of CAD 30 000 (USD 29 115). For the duration of the investment, the conditions which made the young person eligible for a loan must be maintained, and additional requirements may even be added. The maximum amortization time of a loan is 15 years.

FIRA has already directly contributed to the implementation of 15 projects ranging from dairy farming to beekeeping and organic vegetable production. When fully in place, FIRA hopes to contribute to the establishment of about 50 farming operations a year.

(Case study drafted by C. Goemans and M. Graf, adapted from Survey)

Information provided by Ms Magali Delomier, Director-General of FRAQ
More information available at:
www.lefira.ca
www.upa.qc.ca
www.gouv.qc.ca

---

[31] These are subordinated loans. Through this mechanism, the value of the available seed money can be doubled. Up to a maximum of three years, the capital and the interests of the loan do not have to be paid back in order to give the farmer the opportunity to have working capital and set up his business.

[32] FIRA can become a shareholder of the agricultural business through the investment of share capital, which facilitates the transfer of the business.

[33] FIRA can buy land and then rent it to young farmers for a maximum of 15 years. During this period, the land can only be sold to the young farmer renting the land.

## 19. YOUTH VENTURE CAPITAL FUND

UGANDA

In 2011, the Government of Uganda in partnership with DFCU Bank, Stanbic Bank and Centenary Bank launched a venture capital fund of approximately USD 9 800 000 (UGX 25 000 000 000) to start the Youth Venture Capital Fund. The fund was put in place to provide venture capital debt finance to viable projects proposed by young entrepreneurs and to enable them to benefit from associated mentoring services from the participating banks. The fund aimed to support the growth of viable and sustainable SMEs (small and medium enterprises) in the private sector.

To qualify for loans, youth entrepreneurs must be aged between 18 and 35. Each business project must demonstrate its ability to provide employment to at least four people by the end of the loan period and each borrower must present at least two credible guarantors. Eligible sectors include: agroprocessing, primary agriculture, fisheries, livestock, manufacturing, health, transport, education, ICT and tourism. Successful applicants must be willing to receive advice and be ready to participate in financial skills training and mentoring for proper business management. Applicants must be able to prove they are Ugandans and should be able to prove their credit worthiness by presenting a credit reference bureau card. Eligible borrowers must be registered and licensed and they need to have been operative for a minimum period of three months.

Loan amounts are up to ten times the existing value of the business including owners' cash. The loan amounts must be between USD 39 and 2 000 for individuals, while for companies, amounts range from a minimum of USD 195 to a maximum of USD 9 700. The maximum tenure for the loan is four years inclusive of a maximum one-year grace period. Interest rates under this facility are fixed for the duration of the loan. Every year the Government and participating banks review and agree on an applicable rate for all loans to be approved in the subsequent 12 months. A once-only fee (1 percent of the approved loan amount) is chargeable, but payable only on approval of the loan. Personal guarantees of the eligible borrowers and the assets of the borrowing enterprise serve as security. Each borrower benefits from the fund only once.

To date about USD 3 212 851 have been disbursed to 3 000 young people.[34] An additional USD 1 305 220 were added to the scheme in 2012, and a new graduate fund was set up called the Graduate Venture Capital Fund with an allocation of USD 6 425 702 to be implemented with participating financial institutions.

(Case study drafted by F. Dalla Valle, adapted from the Internet)

## 20. YOUTH SOCIO-ECONOMIC EMPOWERMENT SERVICE

REPUBLIC OF MOLDOVA

Between 2005 and 2008, the Youth Socio-Economic Empowerment Project Phase I (YSEEP) was carried out, with funding from the World Bank and UNICEF in 17 regions of Moldova.[35] YSEEP provided technical and financial support to rural and peri-urban young women and men (aged 18–30) for business development (World Bank/UNICEF/USAID, 2009). The main goal

---

[34]    From the Finance Minister, Maria Kiwanuka, in her 2012–13 budget speech to Parliament on 14 June 2012.
[35]    Drochia, Edinet, Donduseni, Falesti, Soroca, Singerei, mun. Balti, Ugheni, Calarasi, Orhei, Nisporeni, Anenii Noi, Criuleni, Cahul, Basarabeasca, Comrat, Cantemir, Causeni.

of YSEEP was to increase the socio-economic empowerment of disadvantaged youth through business creation and participation in innovative community-based service delivery.

Four development agencies (DAs) were contracted to conduct information campaigns, review applications from potential candidates and offer training in "business development", helping participants work out business plans and supporting post-realization. Candidates applying for participation in the programmes needed to fulfil several criteria, including age (18–30), no credit history and residence in a project implementation area.

A total of 800 candidates participated in the DA business development trainings and 215 business plans were formulated, 143 of which were approved for bank loans in addition to WB grants. All the approved business plans were implemented, creating more than 350 new working places (250 for young men and 100 for young women) in a wide range of jobs (63 percent in services, 33 percent in production, 4 percent in trade). The average investment costs to start a new business were around USD 10 000, and financing consisted of 50 percent WB funds, 40 percent bank loans and 10 percent beneficiaries' contributions. Training and assistance (for marketing, accounting etc.) were provided throughout the process and also after implementation.

Given the positive outcome of YSEEP, it was decided to launch Business Development for Youth Economic Empowerment in Moldova (phase II) in 2008. The main funders were USAID (United States Agency for International Development), WB and UNICEF, and the goal was the creation of 60 youth-owned businesses all over Moldova. The capacity development needs of selected young entrepreneurs were first assessed, and on the basis of this assessment, a tailored training curriculum was developed. In addition, field trips were undertaken to learn from best practices. The participants received training in financial management (accounting and financial reporting, taxes, preventive measures), marketing (work with customers, promotional activities), human resource management (employment, on-the- job training) and in specific practical approaches of business development, in both agribusiness and non-agricultural production.

The approach was successful thanks to the combination of non-financial (training and advice) and financial support to young people, allowing them to mature their idea and find the means to implement it.

(Case study drafted by M. Graf, adapted from Survey)

More information available at:
http://www.capmu.md/
http://www.capmu.md/images/docs/Reports/2010/CAPMU%20Progress_Report_2010_ENG.pdf

## 21. FINANCIAL SERVICES FOR YOUTH THROUGH RURAL ENTREPRENEURSHIP

SENEGAL

The young beneficiaries of the Project for the Promotion of Rural Entrepreneurship II (PROMER II) in Senegal had difficulty accessing finance through banks and decentralized financial institutions to support their micro- and small rural enterprises. In order to overcome the challenges, PROMER II has been providing solutions through its Support Service to Rural Finance (SAFIR) component in the regions of Kaolack, Thiès, Fatick, Tambacounda, Kédougou, Kaffrine and Kolda since 2010. PROMER II has identified five financial service providers that are suitable for their rural clients. The West African Development Bank (BOAD), a PROMER II project partner, has provided

FCFA 1 million (USD 2 000) to those five financial institutions to be used as a credit line for the beneficiaries of PROMER II.

PROMER II organizes information sessions to brief rural youth on SAFIR and the different financial service providers, their services and the conditions of access. The youth selected by the project are then trained and coached to develop their project idea and to draft a business plan. The business plan should incorporate all their needs ranging from technical training to financial services and monitoring activities. Subsequently, the youth are put in touch with one of the five FSPs, and additional support is provided to further detail their business plan, focusing on the analysis of their capacity to repay the loan based on the profitability of their activities. The loan aims to cover the working capital as well as the equipment needed for the activities. The project also set up an inventive tripartite guarantee system involving the guarantee groups, the financial institutions and PROMER II. Youth with the same type of micro-/small rural enterprise and engaging in the same type of activities (e.g. the transformation of cashew nuts) are clustered into guarantee groups in order to sign an agreement with the financial institution to receive a loan. The idea behind the group is that the members motivate and help each other to pay back the loan. They also set up a guarantee fund in case a member fails to repay. Clustering groups has proved more effective with women than with men, because women are more likely to motivate their group members to repay the loan. The financial institutions know this and they are thus more likely to trust a women's group than a men's group. The project also acts as guarantor and commits itself to reimburse part of any unpaid debts. FSPs offer loans at a low interest rate and accept this special guarantee system. Youth who take out a loan are accompanied throughout the process, during their activities until repayment of the loan.

Around 33 percent of the beneficiaries of the PROMER II project should be youth and the SAFIR staff have received specific training on targeting youth. As a result, the SAFIR component has enabled 500 young people to join financial institutions, and more than 340 of them have accessed finance for the creation of their businesses.

(Case study drafted by C. Goemans, adapted from Survey)

Information provided by Mr Ameth Hady Seydi, Manager of the SAFIR Component of the PROMER II Project.
More information available at:
www.promer-sn.org
IFAD. 2011. *La promotion des micro et petites entreprises rurales: un moyen efficace pour lutter contre la pauvreté en milieu rural.* Livret de capitalisation PROMER II.

## 22. FRIENDS HELP FRIENDS SAVING GROUP

CAMBODIA

The Friends Help Friends saving group (FHF) was launched in November 2009 by a small group from the CAN (Citizen Action Net for Social Development) youth team, led by Mr Kok Tha and based in Phnom Penh, Cambodia. The group began with ten members (three women) and a total start-up capital of about USD 200, and by July 2013, it had about 91 members, approximately 65 percent of whom from rural areas, owning a total capital stock of USD 62 539. FHF is a not-registered and self-reliant saving group.

The concept of saving groups in Cambodia existed prior to 2009; therefore, FHF's founders were able to research the basic concepts, overall principles and guidelines of establishing, managing and

*(17) FHF members. (18) An FHF member. © FHF*

leading the group. Following consultations with external resource persons, they came together to discuss and lay down their own principles. The main objectives of FHF are to: provide financial support to all group members; contribute to improving the living conditions of group members; and build a network of friendship and solidarity. In order to achieve these objectives, five core values were specified during the founding phase: (1) Save–borrow–return; (2) Honesty; (3) Transparency; (4) Mutual respect; and (5) Mutual trust.

FHF's membership is divided into core members and deposit members, and each member has its own account and shares, starting with USD 20 and each month saving between USD 5 and 500 (according to ability). The business management team comprises a chairman, deputy chairman, secretary and cashier.

"Starting small, but thinking big" was the guiding principle of FHF, which grew and improved, thanks to regular and on-time monthly saving, borrowing, and returning; sharing and contribution of ideas and comments; observation, feedback and a willingness to understand the concept of a saving group. FHF's success is due to members' commitment, honesty, trust and strict adherence to FHF's principles. Members borrow money to cover a range of expenses: school and university fees; course material; investments in the small agricultural sector and small businesses; house building and reparation; increased mobility (motorcycles and cars); and assistance to families.

FHF's role extends beyond saving and borrowing to personality development and fostering of self-confidence; improvement of financial prospects; promotion of healthy living; development of resourcefulness; encouragement of social networking; business creation; reduction of youth problems; and social development and welfare. FHF makes a significant contribution to Phnom Penh's society, helping to solve social and youth problems through flexible, affordable, responsible and sustainable financial support. Youth saving groups continue to be established in Cambodia, in particular in Phnom Penh: CAN Youth Team Saving Group (CYTS), Community Management Course Saving Group (CMCSG), Saving for Our Future (SOF), Ladies Saving Group (LSG) and the Youth For Future Saving Group (YFF).

The "Friendship Saving Federation" (FSF) was established in 2010 and current membership comprises FHF, CYTS, CMCSG, LSG, SOF and YFF. The federation is managed by a committee chaired by all member organizations and which meets on a regular basis. There are 219 youth saving groups in the region of Phnom Penh and participating in FSF with a total capital of USD 88 727. The monthly interest reaches USD 1 448, while the monthly saving is around USD 3 207. This ensures the provision of a total loan of USD 93 903.

(Case study drafted by M. Graf, adapted from Survey)

Information provided by Mr Kok Tha, Chairman of the Friends Help Friends Saving Group, Phnom Penh, Cambodia.
More information available at:
https://www.facebook.com/groups/175127332497573/

## 23. LOAN PROJECT FOR YOUNG ENTREPRENEURS

BANGLADESH

The Grameen Bank model – the "bank of the poor" – was initiated by Prof. Mohammed Yunus in Bangladesh. The underlying premise of Grameen is that, in order to emerge from poverty and remove themselves from the influence of usurers and middlemen, landless peasants, including women and youth, need access to credit, without which they cannot be expected to launch their own small enterprises. When lending to groups, the model›s basic philosophy is based on the fact that weaknesses at individual level are overcome by collective responsibility. The coming together of individual members serves a number of purposes, including: education, awareness building, collective bargaining power and peer pressure. When lending to individuals, microloans are given directly to the borrower. Very small loans are given without any collateral and loans are repayable in weekly instalments spread over a year; eligibility for a subsequent loan depends upon repayment of a first loan.

In 2008, the Grameen Bank started a new project: the Youth Entrepreneur Loan Project (YELP). Most of the young, well-educated students who successfully completed their education financed by the higher education loan of Grameen Bank, struggle to find an adequate job upon completion of their studies. In order to promote self-employment, YELP provides loans to youth who show enthusiasm and entrepreneurial thinking and are hard working. They are motivated to create businesses not only for themselves, but also to offer new job opportunities for others. Loans are not only provided to individuals but also to joint ventures. There is no limit to the size of the loan, but young entrepreneurs are advised to start with small amounts which may be increased as their experience grows. In workshops, the young borrowers are taught to be innovative and to take initiatives, to make environmentally friendly choices and to use indigenous materials as well as local infrastructures. In general, a loan is repayable over two years, but this period can vary depending on the project size, the business plan, costs and return. An interest rate of about 20 percent per annum is charged and repayment begins eight weeks after receiving the loan at the latest. In order to reduce investment risks – natural catastrophes, fires, accidents, serious disease, robbery – a risk fund was also created. This fund is used to deduct 1 percent from the 20 percent interest rate on the loan in the case of an unexpected incident. The fund requires an obligatory life insurance payment by the borrower equal to 3 percent of the loaned sum.

Until August 2011, loans were provided to more than 1 432 young entrepreneurs, of which 154 were young women. Businesses were created in trading, computer services and training, clinics and healthcare, poultry, livestock and fisheries, as well as in phone and fax centres and fashion houses. By 2011, loans provided under YELP totalled USD 1.9 million, and the highest contributed loan was USD 0.8 million. The total repayment rate is 99 percent.

(Case study drafted by M. Graf and F. Dalla Valle, adapted from the Internet)

More information available at:
http://www.grameen.com/

## 24. CROWD-FUNDING: THE GOAT DAIRY PROJECT

GRENADA

The Grenada Goat Dairy Project, a non-profit organization dedicated to sustainable agricultural production, launched itself using kickstarter.com, a virtual platform to source crowd-funding (the term given to initiatives that encourage individuals to pledge money, usually via the Internet). Thanks to the online funding campaign, it was possible to set up a school-based dairy facility for youth development.

The Grenada Goat Dairy Project, known locally as the Goat Dairy, was launched in 2008 in the wake of Hurricane Ivan to train local farmers and educate young people about the benefits of goat production as a sustainable business model. It offers trainings in animal husbandry, improved livestock breeding, development of local feedstuffs, and milk and cheese production. The project organizers developed the idea of establishing a goat farm, but the cost of USD 55 000 was beyond the budget of the Goat Dairy, which covers 70 percent of operating costs through production of its chèvre cheese, sold to local restaurants and markets. It was decided to attempt crowd-funding, despite the risks involved. The funding platform chosen applies an all-or-nothing policy: either the goal in pledges is met by the set date or it is not, in which case the investors keep their contributions.

*(19) Goat cheese produced by the Goat Dairy. (20) Goat, Maxine, from the Goat Dairy.*
© Grenada Goat Dairy Project

The team produced a well-designed campaign, including a video explaining the Goat Dairy's work and a range of rewards for people who made pledges: a donation of USD 25 would earn a "thank you" on a postcard designed by a Grenadian artist, a gift of USD 150 would give free classes at a local yoga studio, and pledges of USD 200 would be rewarded by lunch or dinner at one of the island's top restaurants. The campaign was launched in August 2012, and there were just 45 days to meet the set goal. After a good start, donations dwindled and the Goat Dairy organizers turned to social media – Facebook, Twitter and Instagram – and contacted the local media, as well as youth organizations and consulates abroad. "We also secured a few matching fund pledges with donors who promised to match all pledges up to a given amount," said Malaika Brooks-Smith-Lowe, Director of Public Relations. "For instance, we had a local real estate developer promise to match all pledges made within a 72-hour period up to a total of USD 5 000. This created a lot of excitement and allowed people to essentially double the value of their pledge." The Goat Dairy raised more than the target figure, with pledges totalling USD 63 160, as well as an additional USD 2 779 given by donors who were unable to pay by credit card. Even after deductions for Kickstarter administration and payment processing, there was enough money to start building the educational goat farm at a local primary school. Work started on the facility, to be housed at St Patrick›s Anglican Public School, and was due to open in May 2013.

Through both hands-on and classroom activities, students will learn the responsibilities associated with animal care, growing their own food, composting and record-keeping. To expand its impact, the programme will network with other schools so that more children and young adults can be involved.

(Case study drafted by F. Dalla Valle, adapted from CTA)

More information available at:
www.thegoatdairy.org
www.kickstarter.com

## 25. FINANCE AND MENTORSHIP FOR INNOVATIVE YOUNG SOCIAL ENTREPRENEURS

ASIA

The Youth Social Entrepreneur Initiative (YSEI) was founded in 2005 as a Global Knowledge Partnership (GKP) multistakeholder partnership programme between its members – Thai Rural Net (TRN), Mitra and OrphanIT. YSEI believes in providing access to finance and guidance to young people with innovative ideas, commitment and vision for social change, so that they may emerge as social entrepreneurs and create a lasting impact. This social venture initiative provides financial support to youth in Asia. Through its emergence fellowship, YSEI invests in young visionaries with big ideas who need crucial start-up support to turn those ideas into action. Start-up support includes: financing of up to USD 15 000, development knowledge, tools for social entrepreneurship, technical consulting through mentorship and access to diverse networks. On a practical level, YSEI invests in social ventures with blended values (economic, social and environmental), ICTs and a focus on poverty reduction, improving disadvantaged/marginalized groups, environmental protection, gender equality and human rights. All ventures proposed for investment should be led by youth aged between 19 and 30.

To date, youth-led initiatives receiving YSEI support have included ventures in agriculture supporting food and nutrition security. Brinda Ayer, a young social entrepreneur from India, secured funds to start up a school and community horticulture enterprise in Bangalore in order to supplement the national midday meal scheme with an appropriate level of vegetable nutrition so as to improve overall school enrolment and child health in India. The enterprise provides nutritional horticultural supplements produced via horticulture in polyhouses (greenhouses) and kitchen gardens placed in selected schools. Faisal Islam, from Padma Khulna in Bangladesh, used YSEI support to provide marginalized farming communities with access to an agricultural knowledge management system to improve their livelihoods. The system provides innovative knowledge and global best practices to farmers in their local language and enables farming communities to establish better resource management, access relevant information and market their produce. As a result, farmers at grassroots level are able to form links with national and global expertise while emphasizing local development needs. Other youth-led social enterprises have also been able to secure financing through YSEI in India, Bangladesh, the Philippines and Timor-Leste.

(Case study drafted by F. Dalla Valle, adapted from the Internet)

More information available at:
http://www.ysei.org/
http://orphanit.com/ysei.html

# 3.3 Conclusions

When trying to access financial services, youth from across the world face several common challenges:

> restrictions in the legal and regulatory environment;

> lack of specifically tailored financial products;

> limited financial capabilities;

> reluctance of FSPs to work with clients who have limited trading records and security (often the case for rural youth).

Provision of financial services allows youth to improve their livelihoods and accumulate assets in the long term. Appropriate and inclusive financial services can equip youth with the resources and support to become productive and economically active members of their agricultural households and communities, and make the transition from childhood to adulthood. Non-refundable grants, incentives and start-up capital for promoting rural youth entrepreneurship are instruments of critical importance.

In the context of MFIs in the developing world, the experience of the Grameen Bank [case 23] demonstrates that it is possible to lend to the poor, including youth with no land or collaterals of any kind. Youth are given the opportunity to purchase their own tools, equipment or other necessary means and to embark on income-generating ventures which allow them to escape from the vicious circle of low income >> low savings >> low investment >> low income.

**Grouping** in **informal saving clubs** (see the Friends Help Friends saving group in Cambodia, [case 22], can help rural youth to improve their means for generating savings and increasing their borrowing power. Support is offered to young women and men to access finance through group savings and interest, and the scheme also helps youth to build self-confidence and trust in the group – factors which are key to its successful functioning. It provides a first contact with financial culture and can be a step towards the use of traditional financial services.

**Youth-dedicated products** are beginning to be offered by some commercial banks, for example, the partnership between the DFCU Bank, Stanbic Bank and Centenary Bank with the Government of Uganda and its Youth Venture Capital Fund [case 19]. Nevertheless, youth are still generally regarded by FSPs to be an excessively risky client group and, while initiatives of this kind are emerging, they remain insufficient.

**Mentoring programmes** provide significant opportunities, as most FSP managers prefer to see an experienced adult mentoring youth business owners to help them deal with rapidly changing trade markets (Atkinson and Messy, 2012). Appraisals undertaken in Africa showed that MFIs with strong outreach to youth had well-trained loan officers who were capable of analyzing business prospects as part of the loan analysis. Credit repayment rates for youth were in fact better than for the overall portfolio, in part because of the tighter risk control associated with their loans (Atkinson and Messy, 2012). Mentoring is also effective in helping youth gain access to markets [chapter 5, case 38].

**Lowering the risk of lending** to young people can be achieved through various guarantee mechanisms: risk funds (e.g. the Grameen Bank, case 23), guarantee groups (e.g. PROMER II, case 21) and other partnerships. Alternative models for providing youth with enhanced security for FSPs are needed; they require political will and more partnerships.

**Start-up funding opportunities** exist so that a rural enterprise may access the market, as the example in Benin [case 39] illustrates. However, funding tends to be more readily available to educated youth who are familiar with ICT tools and preferably know a foreign language. The start-up fund for agricultural activities in Canada [case 18] requires the first repayment after three years. This reduces the pressure on youth starting a business and gives them time to get established – a lesson that would also prove useful in the context of developing countries. ICTs also provide access to e-banking and crowd-funding. Crowd-funding sites were successfully used to set up the Grenada Goat Dairy project [case 24]. Such sites offer great potential to get a project started in today's world.

**Competitions** are another potential source of funds, particularly those targeting rural youth, where a good business plan is evaluated and rewarded. As with MFarm in Kenya [case 33], these competitions provide the winners with increased visibility, giving a crucial boost to their business.

A large number of NGOs act as FSPs for small enterprises in poor rural and urban communities, and many target youth in particular (Dalla Valle, 2012). In response to the limited financial literacy (e.g. unfamiliarity with drafting a business plan) among rural youth – particularly those in developing countries – many NGOs offer technical training related to the loan activity, support to new and existing businesses and/or programmes to improve the living environment with the objective of alleviating poverty. YSEEP in Moldova [case 20] successfully combined training with the provision of loans to launch businesses, and also managed to support young entrepreneurs with follow-up training in marketing and management. Rural youth require a wide range of financial services, not only loans or savings opportunities. The above-mentioned YSEEP and the Grameen Bank [case 23] not only provide loans, but also organize training courses on financial literacy, with support to the creation of local businesses.

The research carried out for this publication reveals that, in developing countries, a number of initiatives providing youth with access to finance are still managed or initiated by development partners. This means that they may depend on donors for funding – a dynamic that cannot be sustained in the long run. It is therefore crucial to include governments and other national organizations in the planning process in order to ensure long-term sustainability. Governments, national financial institutions and other national organizations, as well as the private sector, have a vital role to play in the sustainability of youth-inclusive financial services, building on initiatives such as those reported in this publication. Regulatory barriers (e.g. age restrictions) to accessing loans can be overcome by creating a youth-friendly regulatory environment, acting inclusively and protectively towards youth. Such policies could encourage the design of suitable financial services for youth as well as the provision of low-cost delivery channels like mobile and school banking programmes (UNCDF, 2012).

Youth represent the next wave of clients for FSPs and now is the time to educate and include them. If youth are to benefit from an inclusive financial sector, political commitment is required, in addition to coordinating efforts among different regulatory bodies (ministries of education, agriculture, youth, finance, employment and trade), producers' organizations, FSPs, other youth stakeholders and youth themselves.

# 4.  Access to green jobs

### MAIN AUTHOR: TAMARA VAN'T WOUT

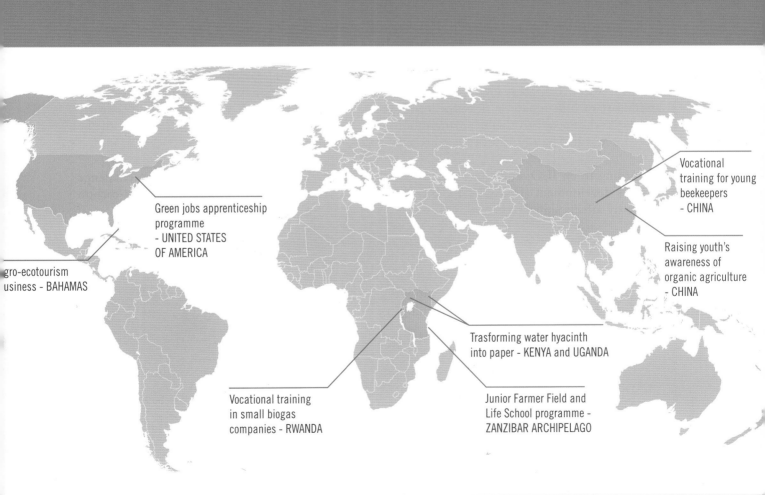

Green jobs apprenticeship
programme
- UNITED STATES
OF AMERICA

gro-ecotourism
usiness - BAHAMAS

Vocational
training for young
beekeepers
- CHINA

Raising youth's
awareness of
organic agriculture
- CHINA

Trasforming water hyacinth
into paper - KENYA and UGANDA

Vocational training
in small biogas
companies - RWANDA

Junior Farmer Field and
Life School programme -
ZANZIBAR ARCHIPELAGO

# 4.1 Introduction

The transition to a green economy[36] not only promotes environmental sustainability, but is also expected to create green jobs in all economic sectors, including agriculture, forestry and fisheries (FAO, 2010). The United Nations Environment Programme (UNEP) defines a green job as "work in agricultural, manufacturing, research and development (R&D), administrative, and service activities that contribute(s) substantially to preserving or restoring environmental quality". It includes jobs that "reduce energy, materials, and water consumption through high efficiency strategies; de-carbonize the economy; and minimize or altogether avoid generation of all forms of waste and pollution" (UNEP, 2008).

A large number of green jobs have already been created worldwide.[37] The creation of green jobs provides potential opportunities for youth in both developed and developing countries to participate in greener economies. Nevertheless, youth's willingness to become involved in these green economies does not necessarily translate immediately into employment. The challenges that youth encounter when they want to access green jobs are discussed in this chapter alongside some solutions that can help youth from around the world to address these challenges.

The box below provides some examples of agriculture-related green jobs, although the list is by no means exhaustive. Green jobs should not only help to reduce the negative environmental impact of economic activity; they need to be decent.[38]

In general, sustainable farming practices, such as organic farming, require more labour inputs as they are relatively labour-intensive compared to conventional farming and have the potential to generate higher social and economic returns (UNDESA, 2010a). This is particularly important in developing countries because, due to lack of sufficient employment, young people migrate from rural to urban areas in search of work. In many developing countries, young women play an important role in natural resource management, and green jobs will, therefore, be of particular interest and concern to them. Furthermore in the green tourism sector, 60 to 70 percent of the workforce is female and half of them are aged 25 or younger.

According to an ILO study, in order to achieve a green economy, the most important challenge is skills development (ILO, 2009; UNDESA, 2010a). The transition to a green and low-carbon economy affects skills in three ways:

> reduced demand for some jobs (skills become redundant);

> increased demand for other jobs (skills must be more widely acquired);

> development of new jobs and "greening" of existing jobs (new skills required).

---

[36] UNEP defines a green economy as "an economy that results in improved human well-being and social equity, while significantly reducing environmental risks and ecological scarcities" (UNEP, 2010).

[37] For example, in 2010 in the United States and in Brazil, the number of green jobs was estimated to be 3.1 million and 2.9 million respectively, i.e. about 2.4 and 6.6 percent of formal employment ( ILO/UNEP, 2012).

[38] Decent work is defined by ILO as "opportunities for women and men to obtain decent and productive work in conditions of freedom, equity, security and human dignity, in which women and men have access on equal terms". Decent work refers to work that allows people to earn sufficient incomes from productive work while ensuring the inclusion of a set of social and labour rights and obligations, such as social security and rights and the possibility to voice out and defend collectively workers' interests (ILO, 2011).

## BOX 1
## EXAMPLES OF AGRICULTURE-RELATED GREEN JOB AREAS

> Environmentally friendly food production – organic farming, composting through reusing residues (e.g. crop, livestock and fish waste, and wood residues), beekeeping, water conservation, agroprocessing and agroforestry

> Energy production from renewable sources – production of biogas from animal manure or crop residues

> Landscape maintenance and biodiversity protection – environmental conservation/ protection and sustainable land management

> Climate change and environmental research, development and policy making

> Environmentally friendly activities in the countryside, – eco-/agro- and sustainable tourism, including on-the-farm agroprocessing to be served to clients

It is therefore necessary to develop the appropriate skills or upgrade existing skills, particularly for young men and women who are entering the job market for the first time. Education is required [chapter 1], both formal (classroom-based and vocational training within secondary and post-secondary education) and informal (internship programmes, on-the-job training and increasing exposure and knowledge through, for example, extracurricular activities). Specific economic, education, social and labour policies also need to be in place, as well as specific labour market services and market information [chapter 5]. However, there is no guarantee that young people will find employment or end up in these jobs.

Related challenges include the promotion of green youth entrepreneurship, certification of green business and access to funding [chapter 3].

# 4.2 Case studies

### 26. JUNIOR FARMER FIELD AND LIFE SCHOOL PROGRAMME
ZANZIBAR ARCHIPELAGO

In 2011, FAO, together with the islands' regional authorities, the Ministry of Agriculture and Natural Resources and the Cooperative Union of Zanzibar (CUZA), and in collaboration with the national Agricultural Services Support Programme (ASSP), introduced specific youth-targeted training using Junior Farmer Field and Life Schools (JFFLS) combined with specific agricultural good practices.

The goal of the JFFLS is to empower vulnerable youth, in particular rural youth, and provide them with employment and livelihood options. JFFLS support vocational trainings specifically tailored to rural settings combining employment promotion and access to markets through the facilitation of youth inclusion in producers' organizations, federations and unions and in collaboration with regional authorities. The subjects of the training are chosen together with the youth from a variety of modules and in collaboration with the partners, on the basis of the local needs. The high adaptability of the learning approach enables the modular methodology to address different socio-economic contexts (conflict, post-conflict, in transition, high incidence of unemployment, food insecurity and malnutrition, poverty or HIV [human immunodeficiency virus]) and populations. The JFFLS programme enhances participants' agricultural, life and entrepreneurial skills through various topics, including, agro-ecosystem analysis (AESA), integrated pest management (IPM), agriculture as a business (e.g. entrepreneurship, marketing, accounting, reporting), hygiene and sanitation, nutrition, HIV and AIDS (acquired immunodeficiency syndrome), child labour prevention and personal development. The topics are addressed through small group discussions, observation, role play and experimentation.

The young farmers who participated in the JFFLS in Zanzibar received training in organic agriculture and the use of natural pesticides, such as the neem tree. The use of organic pesticides and other organic farming practices (e.g. composting) both helped the young farmers reduce the costs of inputs and benefited the surrounding environment and natural resources. The young trainees returned to their communities, where they cost-effectively retrained their young peers and raised awareness of organic agriculture. They gained access to local markets for their organic produce, mainly on the main island of Unguja, but also on Pemba. In particular, the booming hotel industry provided a market for their produce, since the majority of guests are European and are increasingly interested in the origin of the products served in the hotels.

(Case study drafted by T. van't Wout, adapted from information provided by F. Dalla Valle)

More information available at:
http://www.fao-ilo.org/?id=20904
http://www.fao.org/docrep/010/a1111e/a1111e00.htm
Dalla Valle, F. 2013 (forthcoming). FAO, private and public sectors' rural youth employment integrated model – Experiences from Malawi, Tanzania mainland and Zanzibar archipelago – FAO's promotion of decent employment opportunities for rural youth producers in Zanzibar. FAO (available at http://www.fao-ilo.org/news-ilo/detail-fr/fr/c/143823/?no_cache=1).

## 27. VOCATIONAL TRAINING IN SMALL BIOGAS COMPANIES

RWANDA

Given the shortage of skilled labour and the high unemployment, in particular among youth, the Government of Rwanda invests heavily in vocational training development to increase skills, enhance employability and increase the capacity to start up small businesses. The National Domestic Biogas Programme (NDBP) was established in 2007 to help develop a commercially viable domestic biogas sector and contribute to the well-being of rural families, while reducing the pressure on natural resources. A significant component in the project is the training of masons and supervisors of small biogas companies. In order to build quality biodigesters, masons require the appropriate knowledge and skills, because biodigester performance depends on size, location and the quality of the construction materials and appliances. The mason must also adhere strictly to construction norms and is responsible for the operational monitoring and maintenance of the plant.[39] Building good quality biodigesters also requires effective installation supervision and post-installation activities by well-trained supervisors.

The NDBP was developed by the Ministry of Infrastructure with support from the Netherlands Development Organisation (SNV) during the first phase (2007–2011) of its implementation. Since 2012, it has been funded entirely by the Government of Rwanda. With regards to vocational skills development, the Rwanda Workforce Development Authority (WDA),[40] its provincial Integrated Polytechnic Regional Centres (IPRCs), their partners (Japan International Cooperation Agency [JICA] for IPRC North) and the districts joined the NDBP to support the development of a specific training programme for the biogas sector. Its objectives are to increase the private sector's involvement and create jobs for people in rural areas. To ensure sustainability, existing technical and vocational education and training (TVET) institutes, such as the Tumba College of Technology/IPRC North and Kicukiro College of Technology/IPRC Kigali, have been involved to deliver the technical training.

The training lasts four weeks in total, comprising three weeks of practical on-the-job training and one week of theory. Topics taught include general biogas theory, biofertilization, construction, operation and maintenance of biodigesters, as well as business development and marketing. To date, around 360 masons and 150 supervisors have been trained, 80 percent of whom aged 24–35 and 20 percent 35–40. Although the construction sector is male dominated, six female masons and ten supervisors have been trained and two women have successfully operated their own biogas companies under the scheme. The programme has provided services to 2 600 families (11 180 household members), 51 percent of whom are women.

(Case study drafted by T. van't Wout, adapted from the Internet and information provided by SNV)

Information provided by Mr Anaclet Ndahimana.
More information available at:
http://www.snvworld.org/sites/www.snvworld.org/files/publications/the_rwanda_domestic_biogas.pdf

## 28. VOCATIONAL TRAINING FOR YOUNG BEEKEEPERS

CHINA

The Eastern Tibet Training Institute (ETTI) is a non-profit vocational training centre based in Diqing Tibetan Autonomous Prefecture, Yunnan Province of China. The training institute was established in 2005 with the aim of improving livelihoods in Eastern Tibet through vocational training of youth aged 18–40 (the majority of whom in their 20s and 30s).

Since 2010, it implements the Advanced Beekeeping Enterprise Development (ABED) training programme in partnership with the Apiary Research Center of Yunnan Agricultural University. The training programme helps young farmers resident in the Tibetan area to develop honey-making enterprises to generate additional income. Given the fragile upland environment of the

---

[39]   A new design – RW (Rwanda) – has been introduced. It uses burnt bricks or stones, depending on availability, allowing a reduction in costs and therefore increasing access for rural farmers. Within the biogas programme, there are 4-, 6-, 8- and 10-m³ plants. The cost for the cheapest digester (4 m3) is RWF 530 000 (approximately USD 816). The Government provides a flat subsidy of RWF 300 000 (USD 462) for any domestic biogas installation.

[40]   WDA has the mandate to coordinate, regulate and supervise the implementation of an open-ended TVET system in the country. Cabinet formally approved the establishment of WDA on 18 January 2008. WDA operates through five decentralized IPRCs, one in each province. IPRCs run the 25 public vocational training schools, and supervise 150 private vocational training schools. Currently, three IPRCs are fully operational.

Tibetan Plateau, farmers find it difficult to increase their income by expanding agriculture or animal husbandry. The production of honey offers an opportunity to provide an environmentally friendly and sustainable source of income. In the last few years, the demand for honey has increased markedly and the honey from the Tibetan region is renowned for its unique flavour. Although Tibetan farmers have produced honey for their own consumption for generations, only a few have the technical skills to produce honey for the market.

In the three years that the programme has been running, 308 local beekeepers have benefited from advanced training in hive management techniques and have received assistance to establish honey enterprises. Skilled local beekeepers provide ongoing assistance to the participants through beekeeper mentor groups, contributing to the sustainability of the project. The programme has seen increases in yield (by 400 percent) and quality, as a result of the introduction of new techniques and technologies. The honey provides a valuable source of income for farmers in remote areas.

ETTI also runs the Green Technology and Eco-Tourism (GTET) training programme, established in 2011. It enhances participants' knowledge of ecotourism, with case studies in ecotourism development and hands-on skills training in green technologies – the construction of a composting ecotoilet was one element of the programme. To date, 75 youth have been trained in green technologies and various ecotourism methods.

(Case study drafted by T. van't Wout, adapted from the Internet)

Additional information provided by Mr Ben Hillman, Chair of ETTI's board.
More information available at:
http://www.etti.org.cn/rural-enterprise-training/advanced-beekeeping-enterprise-development-program-2012/
http://www.etti.org.cn/wp-content/uploads/2012/08/ETTI%20Annual%20Report%202011.pdf

## 29. GREEN JOBS APPRENTICESHIP PROGRAMME

### UNITED STATES OF AMERICA

YouthWorks, an American non-profit organization in Santa Fe, New Mexico, received funding from the City of Santa Fe Economic Development Division to establish the Green Collar Jobs Apprenticeship Program. It has developed community-based partnerships with Santa Fe Community College, local businesses and organizations, and offers vulnerable youth training in order to develop their skills, raise awareness of environmental issues and increase employment opportunities.

The apprenticeship targets youth in the Santa Fe area, between the ages of 16 and 25 and with problems finding work, due to a criminal record or lack of high school diploma. Youth involved in the organization's other projects – for example, the River Restoration Project and Santa Fe YouthBuild – may also participate. The apprenticeship lasts three months and apprentices receive a salary, subsidized by YouthWorks.

Participants are placed with employers operating in the green economy and offering hands-on and on-the-job experience in, for example, landscaping, horticulture and water conservation. Key issues regarding sustainability and the emerging green economy are introduced. There is a particular focus on the potential opportunities for green jobs within the water sector, related to river restoration, water management and water conservation.

YouthWorks also provides assistance with resumes, cover-letter writing and interview preparation, helping students find jobs once their apprenticeship ends. To date, 35–40 percent of those who completed the training programme have found jobs. On average, wages start at USD 10.29 per hour: USD 3.79 is paid by the organization and USD 6.50 by the employer for the first three months.

The Green Collar Jobs Apprenticeship Program is promoted mainly by word-of-mouth and to date, applications have outnumbered jobs available. Trainees tend to remain in jobs that are considered to be "green", and some even obtain two-year degrees or attend four-year programmes at universities to obtain further qualifications for specialized green jobs.

(Case study drafted by Tamara van't Wout, adapted from the Internet)

More information available at:
www.pacinst.org/reports/sustainable_water_jobs/youthworks.pdf

## 30. RAISING YOUTH'S AWARENESS OF ORGANIC AGRICULTURE

CHINA

Since 2006, over 200 interns have participated in the Community Supported Agriculture (CSA[41]) internship programme, organized by Partnerships for Community Development (PCD). PCD is an NGO based in Hong Kong and funded by Kadoorie Farm and Botanic Garden (KFBG), which was established in 1956 to provide agricultural aid to farmers. PCD's CSA internship programme focuses in particular on organic agriculture and raises youth awareness of organically grown food, local production for local consumption, sustainable diets, rural-urban linkages and the relationship between consumers and producers. The interns are placed with various NGOs, community-based organizations (CBOs), farmers' organizations and rural communities working on organic agriculture and the development of small-scale CSA in rural areas in Guangxi, Sichuan, Guangdong and Hebei provinces and in the vicinity of Beijing.

Although organic food production has increased rapidly in China over the last 20 years, domestic consumption of organic food is still very low as the majority of organic products are exported. It is estimated that organic food accounts for a mere 0.01 percent of China's total food consumption. The domestic market for organic products is quite small and there is limited understanding of organic agriculture. Furthermore, CSA and smallholder farmers have difficulty building solid links with urban consumers. The internship programme attempts to address these difficulties by raising awareness and increasing knowledge and skills among youth in urban areas, many of whom are originally from rural areas, but moved to the city to complete their university education.

The duration of the internship programmes varies from 10 to 12 months, with the possibility of a 12-month extension, decided on a case-by-case basis. Interns are placed by mutual agreement between PCD, the interns and the host organizations. PCD provides information about the potential hosts, and interviews are conducted between PCD, the hosts and the interns.

---

[41] A CSA consists of members that support a "community" farm, where "growers and consumers support and share risks and benefits of food production". While CSAs come in different forms, a shared commitment to build a more local and equitable farming system is at their core and farmers typically use organic or biodynamic farming techniques (DeMuth, 1993).

A variety of topics are covered in training and include the concepts of CSA, cooperatives and sustainable agriculture, including permaculture and organic agricultural practices. In recent years, PCD has encouraged interns and their host organizations, through the provision of small grants, to develop projects based on their needs. Moreover, some former interns have also received assistance to continue exploring CSA and community development issues outside their original hosting organizations. Some of the ideas supported through small grants include:

> the establishment of a community farm fair at Liuzhou, linking consumers in the community with farmers' networks;

> the provision of training on ecoagriculture to farmers by a former intern;

> a study trip to Hong Kong to learn how social enterprises are run and how to make action plans for crossregional sharing with regards to consumer education and ecobreeding of pigs.

> the provision of around ten thematic scholarships to former interns to further their learning through participation in various activities, including attendance at a training on permaculture, learning ecofarming skills, conducting farm visits and exchanges between smallholder farmers, and organizing consumer educational activities.

*(21) CSA interns. (22) Graduation of the sixth group of CSA interns in July 2012. (23) A CSA intern running an organic produce shop in the local community. (24) Intern, Guan Qi, on a placement in Taiwan to learn about agriculture. © PCD*

Through the internship, participants' awareness of organic farming was raised in urban areas. Approximately 80 percent of the interns trained have remained in the fields of CSA, ecological farming and related issues.

(Case study drafted by T. van't Wout, adapted from Survey)

Information provided by Ms Sherman Tang, Programme Coordinator of Partnerships for Community Development. More information available at: www.pcd.org.hk

## 31. AGRO-ECOTOURISM BUSINESS

BAHAMAS

In 2010, the idea for the Siphiwe Honey Gold Farm and Preserve project was born when Raynard Christopher Burnside (a 29-year-old, unemployed agricultural science teacher) was searching for work. Between job applications and interviews, he provided free tour guides to tourists visiting Rum Cay Island and saved the tips that he received. In addition, he won a cash price of USD 5 000 when he entered the Talent and Innovation Competition of the Americas (TIC Americas),[42] and he invested it in the construction of agro-eco-lodges on 5 ha of family-owned land. In addition, he receives donations, mostly from individuals and long-time residents of Rum Cay Island, who share a common interest and have a passion for conserving the environment and promoting environmental education.

Burnside started his business in January 2012 upon completion of the online course, "Agro-ecotourism: Basic elements for implementing an innovative tourism project",[43] which enhanced his skills tremendously. He was offered a grant of USD 200 by the course organizers, IICA and OAS (Organization of American States), which covered half the total costs of the course.

Siphiwe Honey Gold Farm and Preserve, established by Burnside, is a green certified agro-ecotourism business promoting environmental education, natural resource conservation and agronomy research through participation in ecotourism activities. It was certified locally through collaboration with the Bahamas Ministry of Tourism, the Bahamas National Trust and the Ministry of the Environment's "going green" Green Certification Programme. Burnside had to follow detailed guidelines and meet certain criteria. He was subject to five visits from inspectors over a total period of about nine months, before finally being awarded green certification. The business strives to achieve a range of ecofriendly objectives, including reducing solid waste; reusing and recycling; increasing energy and water efficiency; reducing the carbon footprint; using biopesticides; and educating employees and customers about its green business efforts.

The agro-eco-lodges are, for example, built to minimize energy and water use by the occupants. Shade trees are located throughout the property and especially around the lodges, making the lodges cooler. The lodges also have energy-efficient heating and cooling systems and low-energy lighting (fluorescent lights throughout the property). There is a solar electric (photovoltaic) farm

---

[42]  http://www.iica.int/Eng/prensa/IICAConexion/IICAConexion2/2012/N07/secundaria3.aspx
      http://www.thebahamasweekly.com/publish/international/Bahamian_agriculture_science_teacher_wins_
      International_Talent_and_Innovation_Competition_TIC_Award_2012_printer.shtml
[43]  http://www.iica.int/eng/prensa/iicaconexion/IICAConexion/2011/N11/secundaria12.aspx

*(25) Burnside demonstrates that growing seeds in seed trays and selling seeds can be a viable business. (26) A Siphiwe Honey Gold Farm and Preserve agro-eco lodge built over an aquatic pond. (27) Burnside sitting next to one of the outdoor clay ovens. © Burnside*

and reserve and a solar (thermal) hot water system; water reduction measures are also in place. Unwanted food – raw peelings and stems, rotten fresh fruits and vegetables and leftover cooked foods – are used to make compost that is used as fertilizer for crop and vegetable production. Old tyres are reused and turned into feeders for wild animals, or stacked on top of each other for the cultivation of root crops, such as potatoes, yam, cassava, ginger and sweet potatoes, in small areas. Ecofriendly landscaping methods are adopted with no use of chemical fertilizers, pesticides or herbicides, to avoid chemicals ending up in nearby streams, rivers and the water table, damaging aquatic life and harming the water quality. Furthermore, plants native to the Bahamas are used, as they suit the climate and soil conditions and require minimal watering.

The preserve has a particular focus on youth. To date, over 700 youths have participated in the Youth Volunteerism Program (YVP), Annual Agro-eco-tourism Summer Camp (AASC), Community Youth Outreach Program (CYOP), on-farm stay internships and beekeeping programmes. The young people engaged in these programmes receive a certificate upon successful completion.

Youth have benefited from this project in various ways, becoming sensitized and more aware of sustainable agricultural best practices through agro-environmental workshops and outdoor learning facilities. They are also exposed to youth entrepreneurship through the on-farm stay and the internship programme, are empowered and acquire respect for nature and other people – including the community mentors who serve as role models. They become involved in and active agents of change, inspiring other youth to become agro-eco-environmental stewards. To date, three interns have become agro-entrepreneurs: two have created a bee production farm and one has begun organic citrus farming.

In 2012, Burnside's project won the Eco-Challenge Award at TIC Americas. The competition aimed to create a platform for young entrepreneurs to promote their business ideas and was organized by YABT in cooperation with OAS.

(Case study drafted by T. van't Wout, adapted from Survey)

Information provided by Mr Raynard Burnside, project initiator and manager.
More information available at:
http://siphiwehoneygoldfarm.blogspot.com/

## 32. TRANSFORMING WATER HYACINTH INTO PAPER

KENYA AND UGANDA

If water hyacinth, a water weed, is not controlled it will cover lakes and ponds entirely. This not only affects the water flow, but prevents the sunlight from reaching aquatic plants and depletes the water of oxygen, thereby killing fish. The plants are also a prime habitat for mosquitoes. The plant may be controlled by chemical, physical and biological means; however these methods are expensive and often ineffective.

Michael Otiendo, 32, from Kenya and Robert Atuhaire from Uganda, each transformed an environmental issue into a business opportunity and now use the water weed to make paper.

Michael started his business with just KES 10 000 (USD 120) and the knowledge that he acquired from the Kisumu Innovation Centre Kenya (KICK). KICK provides training opportunities to youth living around Lake Victoria for the production of quality and innovative products made from recycled goods (mostly waste material). Robert, on the other hand, launched his business after graduating from university, where he studied Wood Science and Technologies. He promotes his business through a blog, in face-to-face marketing and by word-of-mouth. Robert also involves the poor people of his community in his business: they go to the lake, pick the hyacinth and sell it to him.

A locally constructed machine and an artisanal process are used to transform the plants into paper. The end products include folders, A4 size printing paper, photo frames, shopping bags and gift bags for ceremonies. The water hyacinth can also be mixed with agricultural residues from wheat straw, banana fibres or sugar cane, or with waste paper collected from universities and business offices, to make different types of paper.

(Case study drafted by T. van't Wout, adapted from the Internet)

More information available at:
http://www.businessdailyafrica.com/Artisan-turns-hyacinth-waste-paper-into-cash-machine-/-/1248928/1425398/-/item/0/-/7jas3y/-/index.html
http://ifad-un.blogspot.it/2011/10/sfrome-new-employment-opportunity-in.html
http://paforenterprises.blogspot.com

# 4.3 Conclusions

There is enormous potential for growth in the creation of new green jobs and for upgrading of existing jobs to become greener, contributing to sustainable development, poverty reduction and better inclusion of young people in society. It is increasingly clear that investments must be made in training and education opportunities so that young people can acquire new skills or upgrade their existing skills in order to have access to "green" jobs.

**Environmental policies** – with regard to, for example, reforestation and conservation – are increasingly prioritized by governments, but labour market regulations and vocational training systems are not necessarily being linked and strengthened. Nevertheless, some governments with adequate fiscal capacities have successfully implemented programmes targeting vulnerable groups, including youth, through green jobs initiatives. Countries such as Australia, China and the Republic of Korea have actively promoted green job creation; in Denmark, Ireland and Switzerland, this was connected with training and retraining activities. Australia went one step further, including the development of education, training and skills for sustainability, creating a culture of innovation and building capacity in order to make the transition to a greener economy.[44]

**Providing youth with opportunities** to gain access to training and education, whether formal or informal, helps them move one step closer to a green job. If youth are not included in TVET programmes to build their skills, they will find it difficult to access green jobs, as they may lack the necessary skills.

The case studies outlined in this chapter have shown that there are different ways in which youth can obtain access to knowledge and information to acquire or upgrade skills.

> **Formal education** gives students increasing exposure to education focusing on environmental awareness, pollution and waste management, integrated into subjects such as history, geography, biology and chemical science. At higher levels of education, students can study topics and courses related to organic farming, environmental science and renewable energy within the context of sustainable development.

> **Informal education** activities can permit youth to increase their knowledge to acquire and upgrade their skills. In China, for example, the CSA internship programme organized by PCD allowed young people to gain first-hand experience regarding how to grow organic produce, working alongside farmers and NGOs striving to increase awareness of organic farming among consumers [case 33]. In Tibet, ETTI [case 28] provides an example of a successful vocational training programme on beekeeping: vulnerable youth are given the opportunity to develop their skills, increase their yields, improve the quality of their honey and market their products.

> **On-the-job training** is provided in various cases. On the Siphiwe Honey Gold Farm and Preserve, youth receive training in how to run an ecotourism business and learn practices related to waste management, water conservation and beekeeping [case 31]. In the Green Collar Jobs Apprenticeship Program, run by YouthWorks, youth receive a wage during their apprenticeship, as well as support to find work upon completion of the programme [case 29]. An example of vocational training focusing on upgrading existing skills is the training of

---

[44]  http://apskills.ilo.org/resources/skills-for-green-jobs-consolidated-response

masons and supervisors in small biogas companies in Rwanda [case 27], which results in the building, installation, maintenance and monitoring of high quality biodigesters.

> **Non-profit organizations**, such as EcoVentures International, provide education and training programmes to youth in Mexico and Haiti, focusing on promoting agriculture as a business through the "AgriPlanner simulation" and the "AgriMarket simulation".[45] The JFFLS programme [case 26]promotes agriculture as a business and organic farming to young people, including out-of-school youth, through learning-by-doing and experimental learning.

> **The Internet** provides numerous opportunities for absorbing knowledge and information and following training programmes. Information is available in online courses on sustainable farming offered by various universities throughout the world, and in courses offered by non-profit organizations, such as the Center for Sustainable Development.[46]

In 2009, a youth-led non- profit organization in Nepal, "Team for Nature and Wildlife" established a project, "Youth employment for Green Jobs", funded by the UN-Habitat Youth Fund. It targeted unemployed youth (men and women), aged 18–35, who underwent 18 months of training. A total of 100 young people were selected out of 167 applicants, and at least 75 percent of the project beneficiaries were unemployed youth from slum areas. Following training, they are now technically equipped to produce candles, bio-briquettes, organic compost and organic honey, and they have acquired entrepreneurial skills. The project has encouraged unemployed youth to start their own green businesses: some have started vermicompost production, others are engaged in commercial organic vegetable farming and beekeeping.[47] Burnside's ecotourism business is a good example of how youth can establish a green certified business [case 31], while the green business concept of transforming water hyacinth, a water weed, into paper [case 32] is an example of creative innovation.

Entrepreneurship is central to job creation: involving youth in green enterprises helps improve their access to green jobs. In order to start a business, youth require additional training. In 2011, the Youth Entrepreneurship Facility (YEF) partnered with Inoorero University in the United Republic of Tanzania to promote green entrepreneurship through the development of a postgraduate course on green business (ILO, 2012a). Youth must also find ways to access finance in order to start their own enterprises. The case studies demonstrate a variety of ways in which youth can acquire access to credit, for example: entering youth business competitions; acquiring loans from the Youth Enterprise Development Fund (YEDF), established by the Kenyan Government which provides loans to support green businesses and youth's green business ideas (ILO, 2012a); or from the Indonesian Green Enterprise programme, a collaboration between ILO and the Central Bank Indonesia aimed at creating new green entrepreneurs and green jobs, and enhancing training and coaching as well as promoting green policies at the national, provincial and local levels.

---

[45]    http://eco-ventures.org/folio/curriculum-training/
[46]    http://www.csd-i.org/about-us/
[47]    http://www.tnwnepal.org/download/YouthEmpUNReport.pdf

# 5. Access to markets

MAIN AUTHORS: ALESSANDRA GIULIANI

AND FRANCESCA DALLA VALLE

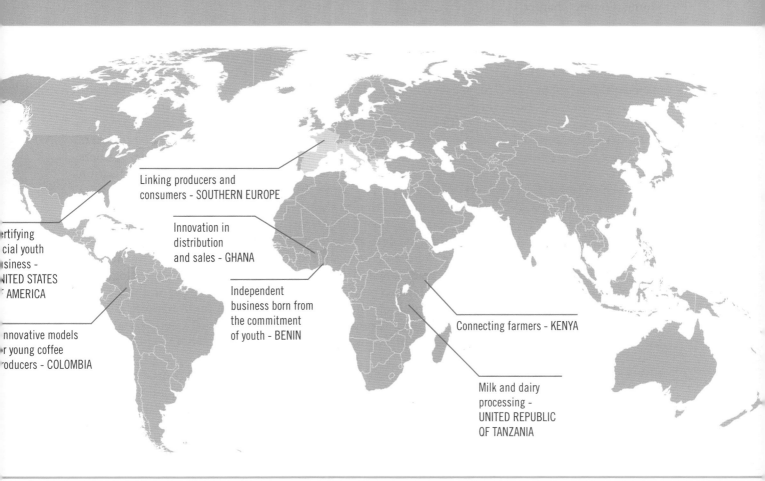

Linking producers and
consumers - SOUTHERN EUROPE

Innovation in
distribution
and sales - GHANA

rtifying
cial youth
siness -
NITED STATES
F AMERICA

Independent
business born from
the commitment
of youth - BENIN

Connecting farmers - KENYA

nnovative models
r young coffee
roducers - COLOMBIA

Milk and dairy
processing -
UNITED REPUBLIC
OF TANZANIA

# 5.1 Introduction

Market access for farmers means the ability to acquire farm inputs and farm services, and the capability to deliver agricultural produce to buyers (IFAD, 2010a). Markets provide the opportunity to generate income, contributing to a reduction in poverty and hunger in developing countries. Markets also drive production to meet consumer demand in terms of quantity and quality (van Schalkwyk *et al.*, 2012). Sustainable access to markets is required to guarantee smallholders an increase in income and to lift them out of poverty.

Since rural youth are the future of the agricultural sector (MIJARC/IFAD/FAO, 2012), their access to markets is vital for boosting productivity, increasing incomes and reducing poverty and hunger for the years to come. Nevertheless, young people face a number of challenges while trying to access markets, even beyond the constraints faced by smallholder farmers in general, in particular in developing countries.

Before accessing markets, young rural people have already faced numerous constraints to starting their farming activities, including difficulties accessing land [chapter 2], agricultural inputs and financial services [chapter 3]. Furthermore, many young people lack experience and knowledge of how markets work; they often lack business, management and entrepreneurial skills, and like many other smallholder farmers, they lack information about prices. Access to information and education is poorer in rural than in urban areas. ICT literacy is also lower, in particular among poor young women [chapter 1].

In the context of booming globalization, the demand for higher value and processed foods, combined with the rise of supermarkets around the world, has implications for the global food marketing system, as it alters procurement systems and introduces new quality and safety standards. Due to increased rural-urban linkages and faster communication, as well as fewer trade barriers, markets are increasingly open and homogenized towards international standards and, as a consequence, more and more competitive. The new procurement systems tend to require large, steady supplies (destined for supermarkets) and favour larger farmers over small-scale producers – which youth often are, particularly in developing countries. These young smallholder farmers are often obliged to maintain compliance with quality standards, cover the costs of certification and invest in technology and infrastructure, as well as in a more skilled labour force. Small young agricultural producers in developing countries can, in theory, sell their products to various kinds of markets: local (rural), emerging urban, regional and international. Improved access to national, regional and international markets is important to allow them to sell more produce at higher prices. However, it can be difficult for them to keep up with the required standards, volume, quality and diversity of goods. Local markets are traditionally the most accessible given the simpler logistics, smaller scale and relatively more moderate competition compared with larger domestic and international markets. However, as distribution channels shift from small local markets towards supermarkets, national and local markets are beginning to imitate international standards, and the market access challenges faced by small young producers no longer apply solely to exports.

In a typical rural market organization, there are a large number of producers and consumers served by relatively few market intermediaries (FAO, 2005). These intermediaries have good knowledge of the market and, for this reason, they can influence market policies. In general, longer marketing chains involving numerous intermediaries present a greater challenge for small young producers, given their age, lack of experience and lack of negotiating power. Young players tend to market their produce through these strong market actors, who frequently take a large part of the profit or supply credit for inputs at high interest rates [chapter 3]. Another reason for youth's lack of market and price information and the asymmetric power distribution (van Schalkwyk *et al.*, 2012) is that they are often not sufficiently well organized (MIJARC/IFAD/FAO, 2012). Greater organization could enhance economies of scale, risk reduction, reduction of transaction costs and access to resources (including funding) (Kruijssen, Keiizer and Giuliani, 2009*)*.

**Young rural women** in developing countries may face additional difficulties in accessing markets, since in many communities their freedom of movement may be restricted because of cultural norms (USAID, 2005) [chapter 1].

In light of the above considerations, it is clear that current market structures are not favourable to young farmers. However, market and infrastructure development in rural areas can lead to employment opportunities for youth in off-farm activities (brokers, intermediaries, market information agents etc.) and in jobs that do not require access to land [chapter 2] or other assets, but that call for specific types of skills and knowledge (e.g. ICT), where youth may have a comparative advantage. Young farmers are generally inclined to be involved in all links of the value chain; they are business oriented and look for innovative ways to make a living as part of a social network – not just in farming.

# 5.2 Case studies

## 33. CONNECTING FARMERS

KENYA

In Kenya, as well as in many developing countries, many smallholders have difficulty buying good quality affordable inputs and selling their products in steady quantities at a fair price. Middlemen have access to market information and are in a position to determine the price to offer for produce.

Three young women entrepreneurs, in their early twenties and members of AkiraChix,[48] – a forum for women interested in information technology – decided to attempt to solve the problem using technology. They founded Mfarm, an agribusiness software company that connects farmers, suppliers and buyers.

Mfarm aims to address the lack of market information via information technology, using mainly mobile phones, since many farmers in Kenya now have access to one. Farmers do not know how much to produce, who will buy, how much and when. Mfarm's formula is simple, and this is part of its attraction to users. A short-code SMS text service supplies farmers with real-time crop prices and market information, connecting them directly to food buyers and, crucially, to each other, so that they can pool their output and access bigger markets. A producer could post an SMS through the Mfarm system to advertise the fact that he or she has, for example, snowpeas to sell, stating the type of produce, quantity, price and location, adopting rapid, text-style language: "sell snowpeas 400 g 4 800 Nyeri".

---

[48]    http://akirachix.com/

As well as connecting producers to the market, the Mfarm SMS service is also available to farmers to check the current prices of different crops, both locally and in other regions, and also the prices of essential inputs (e.g. fertilizers, seeds and pesticides). Mfarm helps farmers find the best deal and obtain bulk discounts through collective buying power. Another service allows producers to enquire about weather forecasts, so that they can plant and harvest their crops with greater confidence.

The young women's efforts to launch the company were given a boost when they won the IPO48 competition to set up a technology start-up in just 48 hours. Beating 87 other contestants, they won a prize of KES 1 million (USD 11 764) to be invested in the business. Less than three years after its launch, Mfarm reaches more than 8 000 farmers in Kenya, contributing to reduced overheads and better market connections. Mfarm has also helped traders, processors, agrifood industries and large retailers obtain reliable and regular supplies and have greater control over the quantity and quality of their purchases.

In October 2013, the company secured a funding of KES 20 million (USD 235 000) from the Safaricom Foundation. This funding will be used to pilot a model to deliver information and better prices to smallholder farmers through mobile technology.

Being young themselves, the three women company founders have a special interest in targeting young farmers. They hope that the use of technology tools, which young people feel comfortable with, will help attract them to farming as a career. "We want to teach youth that farming is not only for the old, because they are leaving everything to the older generation".

(Case study drafted by A. Giuliani, adapted from CTA)

More information available at:
www.mfarm.co.ke

## 34. INNOVATION IN DISTRIBUTION AND SALES

GHANA

Many African countries have traditionally focused on exporting cash crops while domestic markets have been largely neglected and are mostly inefficient. The result is massive food waste and high consumer costs for produce such as fruit and vegetables. Farmers also suffer from low prices and uncertainty over outlets. To face this challenge, a 26-year-old Ghanaian entrepreneur, Richard Ahedor Seshie, founded Vivuus Ltd, a small company designed to improve the collective rural transport system and boost small farmers' income and the income of street vendors by up to 10 percent. Vivuus Ltd. operates out of Ghana and is being launched in neighbouring Côte d'Ivoire, and the model is applicable to sub-Saharan countries in general. As much as 20 percent of food produce distributed to Accra remains unsold and rots, since wholesalers cannot forecast a day's orders and so buy in excess. Ultimately, it is customers who cover the costs and during the last decade, this inefficiency has contributed to price increases of up to 900 percent for certain staples.

Vivuus Ltd operates on a principle of "mobile + mobility" to help informal city vendors of food staples and smallholder farmers escape poverty. It has developed a rural transportation system for the efficient collection of crop surplus, agricultural waste and market residues in urban centres, selling on to third parties or for conversion into biogas and fertilizer. A lead farmer is chosen to

*(28) Explaining the mobile-phone-based sales system m-commerce system to a woman vendor.*
*(29) Vivuus Ltd staff members.* © Vivuus Ltd

oversee collection from other small-farmers and is equipped with a "cargo bike" to facilitate the transport of produce to the market.

Vivuus has also developed a mobile-phone-based sales system enabling women vendors in urban centres to purchase food staples. It sends out field agents to register women vendors, who must have a mobile telephone. Women thus become the main informal vendors, dominating retail sales to consumers on urban markets and improving distribution. Vivuus sends vendors "deal of the day" text messages, offering staples at discounted prices. Thanks to the above-mentioned food collection system, the quantities of food required to meet the increased orders are available.

Vivuus Ltd has a range of supporters and technical partners. It was initially supported by VC4A (Venture Capital for Africa), an Internet platform founded in 2008 to give early stage companies (especially youth-led ones on the continent) effective ways to connect with one another and to secure deal rooms and tools for networking with investors. VC4A is a fast-growing community of dedicated young professionals with a vision to increase Africa's growth through entrepreneurship. The platform has initiated a "matchmaking" mentorship programme called VC4africa.biz that puts young entrepreneurs and young companies in contact with interested investors and clients expanding their marketing opportunities.

(Case study drafted by F. Dalla Valle and A. Giuliani, adapted from CTA)

More information available at:
http://vc4africa.biz/ventures/Vivuus-ghana/
http://vc4africa.biz/blog/2012/08/13/Vivuus-limited-a-holistic-approach-to-accomplish-energy-and-food-security-in-ghana/
https://vc4africa.biz/

## 35. LINKING PRODUCERS AND CONSUMERS

"We Deliver Taste" is a platform aimed at creating a bridge between producers and consumers. Based on the principles of transparency and fair trade, this private registered company focusing on young farmers was created in March 2013 by a group of young people already involved in Slow Food[49] and Slow Food Youth Network.[50] The start-up was achieved thanks entirely to ICT tools (Skype, Facebook, Google Drive and Dropbox).

The platform aims to provide custom-made consultancy to young smallholder farmers to encourage them to respect local traditions, conserve the soils, enhance biodiversity and protect natural resources while producing food products. To improve market access for small-scale young farmers excluded from mainstream supply chains, it helps them create commercial brands. We Deliver Taste then buys their products and places them in high-end markets in Europe and overseas, through a Web shop. We Deliver Taste also organizes gastronomic events and tasting sessions in order to present products to consumers and tell the stories behind their taste. The target group is young and aware consumers. The final goal is to connect them directly with young farmers producing with respect for sustainability, ethics and tradition.

We Deliver Taste work is in Greece, Italy, France, Spain and Cyprus. Outside the Mediterranean area, it has expanded to the Netherlands and the Czech Republic, where, for example, it has successfully marketed "Calypso",[51] a small family brand of olive oil produced in northern Greece and managed by young people.

We Deliver Taste's success is due to a number of factors:

> There is growing consumer interest in the stories and origins of food, not only in Europe, but worldwide.

> Young educated people, who speak a foreign language and have a wide social network, are turning to the land for employment opportunities, pushed by the financial crisis spreading across southern Europe.

> New ICT technologies and the Internet provide excellent opportunities for producers to connect directly with consumers.

> We Deliver Taste's staff have excellent knowledge and expertise, covering a wide range of topics in food, gastronomy and agriculture.

(Case study drafted by A. Giuliani, adapted from Survey)

Information provided by Mr Pavlos Georgiadis, co-founder of We Deliver Taste.
More information available at:
www.wedelivertaste.com
www.facebook.com/wedelivertaste

---

[49]    http://www.slowfood.com/
[50]    http://www.slowfoodyouthnetwork.org/
[51]    www.calypsotree.com

## 36. CERTIFYING SOCIAL YOUTH BUSINESS

UNITED STATES OF AMERICA

YouthTrade was conceived by YES (Youth Entrepreneurship and Sustainability),[52] in partnership with the Conscious Capitalism Institute (CCI).[53] The project was created by young businesspeople and is designed to promote entrepreneurship among under-35s by creating new markets for their products. YouthTrade fosters sustainability and encourages socially responsible investments, taking account of environmental concerns at global and local level. YouthTrade aligns conscious companies with conscious young entrepreneurs. Many of the youth businesses supported by the project are related to agroproducts or agroprocessed products; however, YouthTrade offers support to all kinds of products if they meet the project criteria – i.e. presented by youth, potentially marketable and following a sustainable business model. YouthTrade certifies youth businesses and encourages conscious companies to provide shelf space for youth-produced goods. It also uses social media to urge consumers to shop with a conscious mind-building awareness, establishes corporate buying programmes (YouthTrade Champions) and sets up YouthTrade Clubs to educate the next generation.

To date, YouthTrade has helped over 50 young entrepreneurs under the age of 35 across the United States. It forms partnerships with major companies to distribute and sell YouthTrade-certified products: Nordstrom and also Whole Foods Market, one of the leading foods supermarkets chains selling organic and natural products (rated one of the most socially responsible businesses globally,[54] and ranked high in the US Environmental Protection Agency's [EPA] list of green power partners).[55] In August 2012, YouthTrade developed its partnership with the Babson College, a leading entrepreneurial institute in the United States, to pilot the first ever YouthTrade Innovation Studio; today, YouthTrade is easily accessible to all Babson students and alumni.

YouthTrade has an important role in connecting young organic farmers with young entrepreneurs, giving them the opportunity to learn and grow together. YouthTrade opens up new market channels and the certified young entrepreneurs – who already follow sustainable businesses practices – donate part of their dividends from the enterprise to related social causes. For example, Heidi Ho Veganics, which sells plant-based vegan cheese made from certified sustainable agricultural materials from local producers, donates USD 1 of every product sold to a farm sanctuary. Global Village Fruit, which produces organic jackfruits drinks, uses its profits to fund jackfruit farmers in India so that they may enhance their infrastructure and knowledge.

(Case study drafted by F. Dalla Valle, adapted from the Internet)

More information available at:<br>http://www.youthtrade.com/index.php/en/<br>https://www.facebook.com/IamYouthTrade

---

[52]  http://kelowna.directrouter.com/~yeswebor/index.php/en/
[53]  http://www.consciouscapitalism.org/
[54]  http://online.wsj.com/public/article/SB117019715069692873-92u520ldt3ZTY_ZFX442W76FnfI_20080131.html?mod=blogs
[55]  http://www.epa.gov/greenpower/toplists/top50.htm

## 37. MILK AND DAIRY PROCESSING

Shambani is a processing dairy company based in the Morogoro municipality and established in 2003 by three young graduates from Sokoine University in the United Republic of Tanzania. The founders set up the company following a business course at university and visited various local dairy farms and milk processing industries to learn more about milk processing and its marketing opportunities.

Fresh out of college, they had no collateral and no business record to convince a bank to lend them the money they needed. They began with a single supplier with a capacity of 30 litres and re-invested all their initial profits in the business while increasing their suppliers. The company now receives regular milk supplies from 256 Masaai households, has a processing capacity of 2 000 litres a day and employs 18 full-time staff, as well as generating income for suppliers and a number of part-time workers. The company organizes daily collections in hygienic cooled containers, and transports the milk to the factory, where it is pasteurized and cooled before being processed, packaged and marketed.

Shambani now supplies processed milk and other dairy products to retail outlets within the Morogoro municipality, in the Tanzanian capital Dodoma and in the country's commercial city Dar es Salaam. It produces six different products: pasteurized cultured milk, pasteurized fresh

*(30) Milk collection from farmers. (31)Distribution truck. (32) Processing facilities.* © Shambani

milk, butter, cheese, ghee and flavoured yogurt, which is especially popular with young people in primary and secondary schools and colleges. A door-to-door service has also been introduced to increase the number of customers, targeting busy households in urban areas, where parents have little time for shopping but are keen to give their children good nutritious dairy products. The delivery service uses bicycles and motorcycles, creating job opportunities for young people.

Shambani plans to increase its investment by approximately USD 500 000 in the immediate future, in order that the company may efficiently market its products, improve packaging, expand milk collection and buy new machinery to extend and speed up the processing. The aim is to reach more than 420 small milk producers and increase processing capacity to 4 000 litres a day, resulting in an additional net income of USD 926 per small supplier.

(Case study drafted by F. Dalla Valle, adapted from CTA)

More information available at:
http://www.dairyafrica.com/news.asp?id=40
http://www.youtube.com/watch?v=ofYoslX74sQ
http://dialogue2012.fanrpan.org/newsroom/youth_in_agriculture_award_shambani_graduate_
enterprise
http://www.shambani.co.tz/

## 38. INNOVATIVE MODELS FOR YOUNG COFFEE PRODUCERS

COLOMBIA

In 2006, the National Federation of Coffee Producers (NFCP)[56] in Colombia launched the Innovative Models for Young Coffee Producers initiative, with support from the Inter-American Development Bank and the National Agrarian Bank. The aim of the initiative was to facilitate young coffee producers' access to markets while providing guidance to youth starting and developing their enterprises. The initiative has two major components.

Component one focuses on enabling young producers to set up their own enterprises and identify and evaluate suitable areas for sustainable coffee production. Once the suitable areas are identified, NFCP collaborates with the National Agrarian Bank to organize the purchase of the land. Young producers are then able to begin their own coffee production enterprises. In order to be eligible to apply, potential young producers must have been resident (for at least three years) in the area where the land is located, be aged 18–35 and have completed a minimum of nine years of schooling. In the first two years of the project, 1 300 young producers applied and 225 parcels of land were distributed.

Component two of the initiative focuses directly on the marketing side of the young producers' businesses. Youth are given support to set up collective coffee agribusinesses, known as "coffee business units" (CBUs), while NFCP guarantees the sale of produce at a fair and transparent price through its network of 540 purchase points. NFCP also provides ongoing support through its special advice units, offering regular technical support to ensure that the coffee meets strict quality requirements. NFCP benefits from the arrangement and consolidated its position as the largest

---

[56] The NFCP is a federal and democratic organization representing the interests of more than 563 000 Colombian coffee producers. Through a network of 34 coffee cooperatives with 540 purchase points, the NFCP guarantees all Colombian coffee purchasing of the entire production at a fair and transparent price.

individual exporter of green coffee in the world, representing 27 percent of Colombia's coffee exports.

In the first two years of the initiative, seven CBUs became fully operative, social security coverage (previously non-existent) was introduced and binding community networks were created to foster entrepreneurship, teamwork and build the kind of social capital upon which rural communities depend in order to thrive.

(Case study drafted by F. Dalla Valle and E. Nieto, adapted from the Internet)

## 39. INDEPENDENT BUSINESS BORN FROM THE COMMITMENT OF YOUTH

BENIN

Hervé Nankpan is from a small village in Benin (Glazoué) and a licensed professional in agrofood biotechnology. At the age of 24, he decided to become an entrepreneur – a high-risk sector in Benin – in order to address two major challenges affecting his country: lack of value addition to agricultural products and limited access to knowledge, skills and funding opportunities for starting an independent business, particularly as a young rural entrepreneur.

He wanted to start up an independent business manufacturing cheese and soy sausages, given that soybean is widely produced in Benin and the supply chain is currently booming. He needed a business plan and, therefore, attended the *Programme d'Appui à l'Emploi Indépendant* (PAEI), organized by the *Agence Nationale pour l'Emploi* (ANPE).[57] The training combined theory and practice; he improved the organoleptic characteristics of the cheese he wanted to produce and finalized his business plan. Thanks to his participation in the programme, he also obtained start-up funding from the Fonds National pour la Promotion de l'*Entreprise et de l'Emploi des Jeunes* (FNPEEJ),[58] the Benin national fund for the promotion of youth business and employment.

However, the maximum FNPEEJ financing of FCFA 20 million (USD 41 270) was not sufficient, because the initial investments were very high, because the equipment and machinery needed to be imported. Nankpan then participated in a second programme, *"Talents du monde UEMOA"*,[59] which provided theoretical and practical training in marketing, management, leadership and corporate communication, and comprised an internship in a company corresponding to the participant's own business idea. This training took place in France and lasted three months, after which he returned to Benin, full of enthusiasm and committed to his business idea. He was dismayed when the financial sources he had contacted for funding told him that he was too young to lead a project of more than FCFA 300 million (USD 620 000).

He then came across on the Internet "Entrepreneurs in Africa",[60] a French institution providing technical support to projects in the pre-funding stage. He submitted his business plan but it was not accepted. He then approached "Total Improvement", which studies project feasibility and profitability and helps find funding – his project was accepted. He re-applied to Entrepreneurs in Africa and the project was selected (one of 14 out of 500) for a technical study. Funding was

---

[57]    www.anpe.bj
[58]    https://www.facebook.com/pages/FNPEEJ/163207480436608?sk=info
[59]    Economic and Monetary Union of West Africa (see www.lestalentsdumonde.com).
[60]    www.entrepreneurs-en-afrique.com

therefore made available by both Entrepreneurs in Africa and Total Improvement and he could finally, after four years, start his company.

The Greatcheese Company[61] processes cheese and soy sausage in Cotonou in Benin. The success of this young rural entrepreneur is an example of how the commitment and perseverance typical of young people can be the key to starting up a private business and seizing funding opportunities. A range of trainings and funding opportunities for young entrepreneurs are available on the Internet, implying, however, a high education level, knowledge of a foreign language, ICT literacy and the opportunity to travel and live far from their place of origin – all potential barriers to young rural women and men.

(Case study drafted by A. Giuliani, adapted from Survey)

Information provided by Sourou Hervé Appolinaire Nankpan, owner of Greatcheese Company.
More information available at:
www.agrobenin.com

# 5.3 Conclusions

Access to markets is crucial for young farmers all over the world. In developing countries, it is necessary to enhance productivity, generate increased incomes and thereby reduce poverty and food insecurity. Despite this, most market structures do not favour market access for youth. The case studies provide examples of how youth can increase market access.

To gain access to markets and commence production, access to resources (e.g. land and financial services) is needed. However, access to these resources is often more limited for youth than for older smallholders [chapter 2, chapter 3]. Programmes, initiatives and schemes specifically directed at young people, such as start-up funding opportunities and competitions, can help them overcome these challenges [case 20, case 18, case 23, case 33, case 39]. These case studies show that in both developing and developed countries, innovation, perseverance and commitment are crucial for making the most of these opportunities.

**Education and training** are essential if youth are to seize marketing opportunities and create their own business. Training programmes should be designed to respond to the needs of young producers [chapter 1]. In both the United Republic of Tanzania [case 37] and Benin [case 39] young entrepreneurs fresh from university or training courses used the knowledge and skills acquired to draw up a business plan. A certain level of education may also be a prerequisite when applying for financial resources [case 37, case 39]. Mentoring programmes such as the young coffee producers in Colombia [case 38] offer numerous advantages: knowledge and competencies to comply with market requirements; specific skills development; and increased confidence.

Young farmers are new in the market and have only limited networks and contacts with buyers, which restricts their access to **market information**. However, modern market information services now exist and the development of ICTs facilitates marketing and trading. Youth tend to be adept at learning to use new technologies and they may already use ICT tools for social networking. They

---

[61]    www.agrobenin.com

therefore have a comparative advantage in accessing market information and can overcome the barrier of asymmetric power distribution (MIJARC/IFAD/FAO, 2012).

> The growth in ICT-based market information services allows the various actors in the value chain to connect, for example, MFarm in Kenya [case 33] provides farmers with information regarding prices through SMS services and connects them with traders and consumers. Meanwhile, the mobile-based sales system developed by Vivuus Ltd [case 34] links women retailers in urban markets with producers in Ghana, improving distribution, lowering prices and allowing farmers to sell more.

> ICTs can enhance agricultural extension services [case 7, case 33].

> ICT tools are used to sell products to consumers (e.g. through the Internet). We Deliver Taste in Europe [case 35] provides custom-made online consultancy services to young farmers to tap into niche markets and also uses a "Web shop" to sell food products.

There is increasing evidence that being a member of a (youth) **producers' group**,[62] can help young people overcome the challenge of accessing markets. Organizations can give youth the necessary bargaining power to interact on equal terms with other market actors. Producers' organizations can also help reduce transaction costs as well as realize economies of scale when buying agricultural inputs and selling agricultural produce (IFAD, 2004). Acting collectively may enable youth to deal with transportation and storage issues, acquire technologies and certificates to comply with required quality standards, and reach the necessary scale to supply the desired quantity of their products in order to access larger markets. The National Federation of Coffee Producers in Colombia (NFCP) provides support to create collective coffee agribusinesses (CBUs), guaranteeing a certain purchase of produce at a fair and transparent price [case 38]. Vivuus Ltd. [case 34] created an efficient collective rural transportation system: run by a head farmer – chosen to aggregate produce and provided with a "cargo-bike" – it enables proficient assembly of crops and the collection of agricultural waste residues to be sold or transformed into biogas and fertilizer.

High-value agricultural products and **niche markets** also offer opportunities to young farmers all over the world (Hellin, Lundi and Meijer, 2009). This can be particularly beneficial for young women farming in developing countries, as FAO reports that women play an important role in many of the high-value agricultural market chains in Africa and Latin America (FAO, 2011a). Consumers, mostly from industrialized countries, but also in transition and developing countries, are increasingly concerned about fair trade and interested in organic, natural, healthy, environmentally friendly and traditional products (Fréguin-Gresh, Losch and White, 2010). Agrotourism, culinary tourism, herbal medicine and natural cosmetics are becoming ever more popular, in particular among the young generation, which is rediscovering local tradition and culture as a countertrend to globalization. Young farmers can exploit these rising opportunities thanks to their creativity, flexibility and interest in addressing novel and niche markets. We Deliver Taste, founded by members of the Slow Food Youth Network (SFYN),[63] helps young smallholder farmers in southern Europe produce territorial products (FAO, 2009b) in a sustainable way, with respect for natural resources and traditional knowledge [case 35]. Globally recognized certification schemes – environmental certification, social certification and branding – are another way of

---

62    For example: *Synthesis of deliberations of the Fourth Global Meeting of the Farmers' Forum; and Youth: the future of agricultural cooperatives* (see http://www.fao.org/docrep/017/ap668e/ap668e.pdf).

63    The Slow Food Youth Network (SFYN) is an international network of young people who bring about changes in the field of food production and consumption. It was founded by a number of enthusiastic and motivated young people with a passion for good, clean and fair food, and with an interest in sustainability issues (see http://www. slowfoodyouthnetwork.org/).

accessing markets and promoting products. YouthTrade [case 36], a youth certification scheme created by young entrepreneurs in the United States, demonstrates how a label of sustainability can have an impact on market access. Organic agriculture provides a potential entry point into local markets,[64] as seen in the Zanzibar Archipelago [case 26]. The certification and niche product market – well known in developed countries – is now also emerging in developing countries, but the challenges are greater: the high cost of certification, lack of awareness among consumers and the need for promotion.

---

[64]    http://www.fao-ilo.org/news-ilo/detail/en/c/143823/?no_cache=1

# 6. Engagement in policy dialogue

MAIN AUTHOR: CHARLOTTE GOEMANS

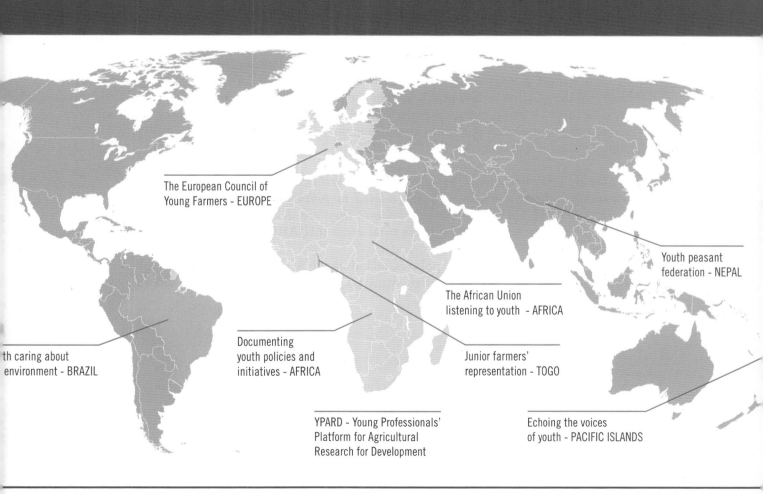

The European Council of
Young Farmers - EUROPE

Youth peasant
federation - NEPAL

The African Union
listening to youth - AFRICA

th caring about
environment - BRAZIL

Documenting
youth policies and
initiatives - AFRICA

Junior farmers'
representation - TOGO

YPARD - Young Professionals'
Platform for Agricultural
Research for Development

Echoing the voices
of youth - PACIFIC ISLANDS

# 6.1 Introduction

Youth have been hit the hardest by the global financial crisis. All over the world, youth unemployment rates are higher than adult unemployment rates (UNDESA, 2012). North Africa and the Near East (with youth unemployment rates of 27.9 and 26.5 percent, respectively) face the world's greatest youth employment challenge (ILO, 2012b). Modern agriculture can offer significant opportunities for job creation for young people (World Bank, 2009). However, in order to attract young people to the sector, an adequate enabling policy and regulatory environment is fundamental (ILO, 2012c).

It is increasingly recognized that youth participation has an important role in decision-making and policy dialogue, and policy-makers are urged to work not only for but with youth. For example, the World Programme of Action for Youth to the Year 2000 and Beyond, which guides the United Nations Youth Agenda, identified *the full and effective participation of youth in the life of society and in decision-making* as one of its ten priority areas (UNDESA, 2010b). The United Nations General Assembly proclaimed the year commencing 12 August 2010 as the "International Year of Youth: Dialogue and Mutual Understanding" (UN, 2010), focusing on three key areas, including "Mobilize and engage (increase youth participation and partnerships)". The recent Arab Spring revolutions have seen youth stand up to claim their right to fully participate in society (Ncube and Anjanwu, 2012; Youth Employment Network, undated).

However, there is still a long way to go to ensure the active participation of youth in policy processes. Too often their participation remains token or passive.[65] Seniority is frequently associated with authority, and youth are not expected or allowed to speak out or voice their concerns, let alone have a role in policy development processes (Lintelo, 2011). In many developing countries, young women's participation in policy-making is particularly challenging as a result of traditional beliefs about the suitability of women to hold decision-making positions and the persistence of gender inequalities at household level (World Bank/IFAD, 2009). Little information is available on the participation of youth in policy processes specifically related to agriculture and rural development. A joint study by MIJARC, FAO and IFAD found that rural youth rarely participate in the formulation of policies concerning them, and rural youth informants from Africa and Latin America stated that:

> they are often not seen as "equal parties" but rather as uninformed, indecisive and trouble makers ... national policies related to youth in agriculture are often not implementable and/or sustainable because they are designed by others who are unaware of the situation of youth in rural areas. (MIJARC/IFAD/FAO, 2012)

Although some legal documents and policies, including the African Youth Charter, explicitly state youth's right to participate in policy design, many young women and men remain unaware of their rights in this regard.[66]

---

[65]  The UN Youth Programme has identified five levels of participation: (1) **Information providing**: youth are informed of the policy and activities selected by decision-makers; (2) **Consulting, decision maker-initiated**: decision-makers decide when and on which topics youth are consulted; (3) **Consulting, youth-initiated**: youth can put subjects forward, but have no decision- making powers; (4) Shared decision-making or co-management: elders and young people share decision-making powers; (5) Autonomy: young people take initiatives and conduct projects themselves.

[66]  African Youth Charter, Art. 14.2 e). Findings from the survey launched for the purpose of this publication.

Furthermore, policies often fail to reflect the heterogeneity of youth and tend to target non-poor men living in urban areas (Bennell, 2007). Consultations are often held in urban areas and in the official language(s) of the country, thus excluding uneducated, rural and poor youth (Lintelo, 2011). There is a lack of comprehensive data on rural youth as a distinct group, resulting in policies that do not respond to the real challenges faced by rural youth (IEG, 2013). Likewise, more detailed research is necessary to document the aspirations of rural youth (Leavy and Smith, 2010).

In order to be able to actively participate in policy dialogue, rural youth need the right skills. Not all rural youth are born leaders, therefore organizations that represent their interests and which can lobby on their behalf have an important role to play. The MIJARC/FAO/IFAD study shows how rural youth are not sufficiently united and they see this lack of unity as a major reason for their limited voice in policy-making processes. There are only a small number of organizations representing only rural youth, and those that do exist often lack financial resources, are rather small and informal, operate at local level, and have little bargaining power in policy processes. Mixed organizations,[67] on the other hand, are often involved in policy dialogue, although there is limited active youth participation in decision-making and leadership. In developing countries, young women face particular constraints to participating in rural organization management (MIJARC/IFAD/FAO, 2012) for a variety of reasons: they generally have lower literacy levels than men; they often lack the confidence to defend their interests; and they have limited mobility and time availability due to the need to combine household duties with a heavy workload (World Bank/FAO/IFAD, 2009).

# 6.2 Case studies

## 40. YOUNG FARMERS' REPRESENTATION

TOGO

When the first Togolese National Farmers Forum (NFF) was held in 2009, one young participant, Eloi Tegba Toi, noticed that almost all the others were over 40 years old, and he began to explore ways to increase youth participation in the agricultural development of Togo. He approached the National Youth Council (CNJ) and with their support managed to organize a meeting with 140 young farmers representing all 35 prefectures of Togo. However, the meeting's recommendation to establish a national network of young farmers was not implemented due to lack of means.

In 2010, when the second NFF was organized, Eloi wrote to the President of Togo and went to the Ministry of Agriculture (accompanied by the president of the CNJ) to request the participation of youth. Eloi's lobbying resulted in the Ministry granting 60 places (out of 1 500) to youth to participate in the NFF. During the forum, youth proposed that a national network of young farmers be set up and requested state support. The President then instructed the Ministry of Agriculture and the Ministry of Local Development, Handicraft, Youth and Youth Employment to provide the necessary support. In July 2010, the network of young producers and agricultural professionals of Togo (REJEPPAT) was successfully created as a youth college within the Togolese Coordination Committee of Farmers Organizations and Agricultural Producers (CTOP). Today, the youth network counts 1 403 member organizations and around 14 751 individual members, of which 5 502 are women from the whole of Togo.

---

[67]    Mixed organizations consist of both young and older members from rural areas.

For the third NFF in 2011, REJEPPAT was invited to be part of the forum's organization committee. A special workshop was organized for rural youth and a stand exhibited the young farmers' agricultural products. REJEPPAT now participates in all national events and meetings related to agriculture, and it is often invited to conferences organized by IFAD, ILO and other international and regional institutions.

As a result of REJEPPAT's lobbying, the State cleared farmland and supported rural youth setting up in farming. REJEPPAT took part in the drafting of the national policy on access to land for youth and women, and facilitated market access by contacting the Togolese Chamber of Agriculture who put them in touch with the International Trade Centre (ITC). ITC financed REJEPPATs prospective market visit to Morocco to identify suitable markets for their pineapples. As a result of these visits, the Togolese national cooperative for the pineapple value chain (COONAFAT) became an important partner of REJEPPAT. REJEPPAT members now occupy important posts within COONAFAT, and the cooperative exports pineapple and its derivative products to Morocco. Thirteen REJEPPAT members were trained by the Songhaï centre in Benin with the support of the State and ROPPA

(33) *Eloi (third from right) and other REJEPPAT members in their pineapple field.* © REJEPPAT

(Network of Farmers' and Agricultural Producers' Organisations of West Africa). Other members were trained in agricultural techniques, agro-entrepreneurship, storage and marketing skills.

REJEPPAT also started its own small livestock project in the Yoto, Vo and Haho prefectures, with the assistance of a specialist trainer. The project aims to stop the seasonal migration of rural youth by providing them with an income during the dry season through the rearing of small livestock, such as chickens, ducks, rabbits and pigs. Staff from the prefectures inform the youth in their area about these trainings, and to date, around 180 young people (54 percent of which young women) have received training in animal husbandry and have received a start-up kit in exchange for a small contribution of about FCFA 10 495 (USD 21). Most now rear their own livestock, and the project has proved so successful that even the elderly are asking REJEPPAT to become members of their network.

(Case study drafted by C. Goemans, adapted from Survey)

Information provided by Mr Toi Eloï Tegba, President of REJEPPAT.
More information available at:
http://www.ctop-opa.org/
http://cnj-togo.asso-web.com/
http://www.intracen.org/
www.songhai.org
www.roppa.info/

## 41. THE AFRICAN UNION LISTENING TO YOUTH

AFRICA

The years 2009–2019 have been declared by the African Union (AU) as a "decade of youth development in Africa" to help speed up the implementation of the African Youth Charter,[68] adopted in 2006, and, more specifically, to support the development and implementation of youth programmes and policies.

The AU immediately launched the policy dialogue concerning youth and development, and in 2010, it chose "Accelerating Youth Empowerment for Sustainable Development" as the theme for its 2011 summit. In preparation for the summit, high-level consultations on youth development issues were organized and the African Youth Forum was held as a pre-summit event. FAO contributed to the forum by providing expertise on rural youth employment and mechanisms to increase youth presence in the agricultural sector. Participants in the forum included young civil society leaders, members of national youth councils, youth members of parliament, and entrepreneurs between the ages of 16 and 29 from all 54 AU Member States. At the end of the forum, resolutions and recommendations were drawn up and submitted to the AU.

During the 2011 AU summit, decisions were made, based on the recommendations provided by the youth representatives:

> *(1) all AU Member States should advance the youth agenda and adopt policies and mechanisms towards the creation of safe, decent and competitive employment opportunities (...); (2) the Commission in collaboration with its partners should elaborate a Technical and Vocational Education and Training (TVET) framework, addressing specifically the domains of Agriculture and ICT (...); (3) Member States*

---

[68] http://www.au.int/en/sites/default/files/AFRICAN_YOUTH_CHARTER.pdf

*should provide to the AU Commission adequate resources for the advancement of the Youth Agenda, including the funding of the Pan African Youth Union; (4) all trained Young Volunteers should be deployed as soon as possible after their training (...); (5) the African Union Commission should lead the organization of a side event on "Accelerating Youth Empowerment for Sustainable Development" during the United Nations High-Level Meeting on Youth in July 2011 at the UN Headquarters (...).*

A large number of youth representatives were present at the summit and included in the discussions.

After the summit, the dialogue continued and representatives of African youth organizations were invited to the Eighth Comprehensive Africa Agriculture Development Programme (CAADP) Partnership Platform Meeting in 2012. Statements underlined the need to focus on youth, in order that they might benefit from the increasing attention being given to the development of African agriculture through the CAADP. A recommendation was drawn up to champion youth participation in agriculture by, for example, launching a robust campaign targeting youth about the implications of engaging in the agricultural sector.

The AU paved the way for youth inclusion in national policy dialogue and it is now up to its Member States to implement all these recommendations and ratify the African Youth Charter. Since the AU youth consultation process, various Member States have started to systematically invite youth to their national policy dialogues.

(Case study drafted by C. Goemans, adapted from information provided by F. Dalla Valle)

Information provided by Ms Francesca Dalla Valle, FAO Youth Employment and Institutional Partnerships Specialist.
More information available at:
www.au.int
www.nepad-caadp.net

## 42. DOCUMENTING YOUTH POLICIES AND INITIATIVES

AFRICA

The Food, Agriculture and Natural Resources Policy Analysis Network (FANRPAN) is a regional policy research and advocacy network which plays a leading role in supporting youth engagement in agricultural policy processes. FANRPAN believes that a lack of evidence-based policies – due, in part, to a failure to consult youth itself – produces responses that are frequently at odds with young people's own aspirations, strategies and activities.

As part of its strategy to involve young people in decision-making and provide them with a platform to voice their concerns, FANRPAN dedicated both its 2011 and 2012 Regional Multistakeholder Food Security Policy Dialogues, held in Swaziland and the United Republic of Tanzania respectively, to considering concrete ways to engage youth in agricultural policy processes. The 2011 dialogue, attended by more than 250 delegates from 23 countries, focused on the theme "Advocating for the active engagement of youth in the agricultural value chains".[69] The gathering gave young people the chance to share their aspirations and express their opinions on

---

[69]    http://dialogue2011.fanrpan.org/

*(34) FANRPAN Regional Multistakeholder Food Security Policy Dialogue 2011, Swaziland.* © FANRPAN

what must change if they are to have an active role in shaping agricultural policy. Six country case studies on youth policies and initiatives were then carried out, with a special focus on agriculture.[70] The studies, conducted by young professionals, took place in Malawi, Mauritius, South Africa, Swaziland, the United Republic of Tanzania and Zimbabwe. The findings were presented at the FANRPAN Regional Multistakeholder Food Security Policy Dialogue in September 2012, where the focus was on moving from ideas to practice in youth policy engagement. In 2013, three more country case studies were commissioned in Lesotho, Mozambique and Zambia. It is hoped that national authorities and partners will use the reports to redirect their efforts and involve youth in formulating public policy and implementing programmes for the agricultural sector. FANRPAN's other activities include: a training session on agriculture value chains and policy advocacy for women farmers and youth in Lesotho; nine internships; the FANRPAN 2013 Youth in Agriculture Award; and the participation of three youths in the Ninth CAADP Partnership Platform.[71]

(Case study drafted by C. Goemans, adapted from CTA)

More information available at:
www.fanrpan.org
http://www.fanrpan.org/projects/youth-in-agriculture/

---

[70]  http://dialogue2012.fanrpan.org/country_case_studies
[71]  http://www.nepad.org/foodsecurity/caad-partnership-platform

## 43. YOUTH PEASANT FEDERATION

In 2007, the All Nepal Peasants' Federation (ANPFa) integrated a separate youth wing: the Youth Peasants Federation (YPF), which launched a national campaign to mark the occasion and now counts over 25 000 active members from 40 different districts. ANPFa includes young leaders in the policy-making processes.

Training courses help develop the leadership skills and professional capacity of the young members, and specialist camps are held on various topics such as climate change and sustainable farming practices. YPF helps its producer members link up with consumers. Another successful initiative has been the creation of cooperatives for the production of milk, coffee and cardamom – and the cooperatives also facilitate youth's access to credit and other services. Youth's agricultural activities have become more profitable, motivating other young people to go into farming. YPF is also a member of IFAD's Farmers' Forum and is involved in IFAD's activities related to the development of farmers' organizations in Nepal.

Pramesh Pokharel, youth leader and member of YPF, believes that YPF's initiatives have given youth a stronger voice and enhanced the lobbying for youth participation in policy formulation and implementation. Various mixed producers' organizations in Nepal have established youth sections and are including more young people in their leadership.

(Case study drafted by C. Goemans, adapted from Survey)

Information provided by Mr Pramesh Pokharel, youth leader and member of YPF.
More information available at:
www.anpfa.org.np
http://www.ifad.org/farmer/index.htm

## 44. ECHOING THE VOICES OF YOUTH

In 2008, the Ministers of Agriculture of the Pacific Islands launched a call to "explore ways in which more young people could be supported to take up careers in agriculture". The Pacific Community (SPC), in partnership with CTA and the Pacific Agricultural and Forestry Policy Network (PAFPNet) responded to the call and carried out a survey in 2009 to explore youth's relationship with farming. The outcomes of the survey have been used to draw up a regional Pacific Youth and Agriculture Strategy 2011–15 "Echoing the voices of Pacific Youth".

Development of the strategy involved various steps. Firstly a survey was carried out and interviews were held in Kiribati, Tonga and Fiji with youth, their parents and their community members. Sessions on youth in agriculture were organized at the Pacific Youth Festival in 2009. National consultations and an online consultation process then took place with the youth and their communities, giving them the opportunity to analyse and check the outcomes of the survey and participate in the development of the strategy. An essay contest was launched as part of the consultation process and SPC-PAFPNET developed a Facebook page for Youths in Agriculture, encouraging young people to discuss policies and strategies. Throughout the development process, awareness was raised on the capabilities, aspirations and needs of young people. It was reported that "the use of questionnaires and presentation of the resulting data to the youth, with the opportunity for them to participate in interpreting their own data, was a new level of engagement and participation".

*(35) A group of young farmers. (36) A young farmer. © SPC*

The IFAD-funded Mainstreaming of Rural Development Innovations (MORDI) programme, which started in 2006 in Fiji, also gave rural youth a voice in their communities. Where youth were once excluded from decision-making at community level, they now work with community leaders and the elderly to elaborate the village development plans. IFAD consulted segregated groups of women, men and youth on community development, and when the outcomes of youth group discussions were presented to the elderly, they realized the importance of including them in the village development committees (IFAD, 2011).

(Case study drafted by C. Goemans, adapted from CTA and the Internet)

More information available at:
www.spc.int/lrd
http://www.facebook.com/pages/Pacific-Youth-in-Agriculture-Network/124321184278078?ref=ts&fref=ts
http://www.facebook.com/spc.int?ref=ts&fref=ts

## 45. YOUTH CARING ABOUT THE ENVIRONMENT

BRAZIL

In 2005, a second National Children's Conference for the Environment was held in Brazil as a follow-up to the 2003 conference. The aim was to facilitate the implementation of Brazil's National Environmental Education Policy,[72] finding opportunities to engage students, teachers, youth and their communities in environmental sustainability.

---

[72] The National Environmental Education Policy is the joint responsibility of the Directorate of Environmental Education of the Minister of Environment and the General Coordination Organisation of Environmental Education of the Ministry of Education.

During the preparation stage, youth and communities were mobilized. Debates took place, bringing together students aged 11–15 from primary schools, indigenous communities, *quilombolas* (an ethnic minority in Brazil), rural settlements and groups of street boys and girls, to discuss issues related to international agreements on biodiversity, climate change, food and nutritional safety, and ethnic and racial diversity. Over 11 475 private, public, urban and rural schools participated in this initiative. Of the 178 participating communities, 40 percent were indigenous communities, 27 percent from rural settlements, 19 percent groups of street boys and girls and 14 percent were *quilombola* communities.

The outcomes of these community-based dialogues were presented at the conference and conference activities were coordinated at three levels: (1) national; (2) decentralized state institutions; and (3) local. Public institutions and civil society collaborated to ensure national mobilization, and over 500 adolescent delegates, 70 facilitators from the youth environmental collectives[73] and 17 young facilitators from other Latin American countries participated, with about 66 percent of the participants being girls. As a result of the conference, a Charter of Responsibilities, "Let's take care of Brazil", was drawn up and presented to the Brazilian President and the Ministers of Education and Environment.

In order to ensure implementation of the Charter, Commissions for the Environment and Quality of Life (COM-VIDAs) were established. Led by the youth environmental collectives, they aim for an ongoing exchange between schools and communities, helping to institutionalize environmental education.

In 2009, a third national conference was held in Brazil to continue the discussions, and in the light of the success of the three national conferences, Brazil decided to organize an international conference to discuss global socio-environmental changes with a focus on climate change. The conference involved 600 young participants from 47 countries and resulted in a Charter of Responsibilities: "Let's take care of the planet".

(Case study drafted by C. Goemans, adapted from the Internet)

More information available at:
http://www.charter-human-responsibilities.net/IMG/pdf/NATIONAL_CONF_PROCESS_ENG_NOV08.pdf
UNICEF. 2012. *Climate change adaptation and disaster risk reduction in the education sector. Resource manual.*
UNICEF. *Process and outcomes of the Second National Children and Youth Conference for the Environment, 2005/2006.*

---

[73] These collectives were set up with the support of the Ministry of Education in the different states of Brazil and bring together youth aged 15–29. The collectives facilitate activities at local level, contribute to the national event and implement the COM-VIDAs. Their work is guided by the following three principles:

1 Young person choosing young person (selection) – Young people are at the centre of decision-making, which is done by young people themselves and not by third parties.

2 Young person educating young person (facilitation) – Young people are social subjects who live, act and intervene in the present, not in the future.

3 One generation learning from another – Intergenerational dialogue, collaboration and action are necessary for sustainable development.

(37) *Joris Baecke (former CEJA President) sitting at the panel with Dacian Ciolos (EU Commissioner for Agriculture and Rural Development) and Simon Coveney (Irish Minister for Agriculture, Food and the Marine) at the European Young Farmer Conference hosted by Macra na Feirme in Dublin, 11–13 March 2013.*

© CEJA

## 46. THE EUROPEAN COUNCIL OF YOUNG FARMERS

EUROPE

The European Council of Young Farmers (CEJA) was created in 1958. Today, CEJA comprises 30 European member organizations from 23 EU Member States, plus one associate member (Croatia). CEJA's main objectives are to: promote a younger and more innovative agricultural sector across the EU; and establish better working and living conditions for young people both setting up in farming and already young farmers. CEJA's cause is highly relevant to today's Europe, considering that only 16 percent of the women and 18 percent of the men working in agriculture in the EU are younger than 35. Only 6 percent of agricultural holders in the EU are below the age of 35, and female farmholders have significantly smaller farms than their male counterparts. Moreover, European farmers under 35 reveal 40 percent more economic potential and 37 percent more utilized agricultural area (EU, 2010; 2012). CEJA provides a platform for dialogue between young farmers and European decision-makers, defending young farmers' interests to European and national institutions. The organization strives for a Common Agricultural Policy (CAP) which prioritizes young farmers and generational renewal in the sector. CEJA is led by an elected president and four vice-presidents and has the support of the Brussels-based secretariat. Many of CEJA's alumni have taken up roles at European level, becoming members of the European Parliament or European Commission officials, and many others successful agro-entrepreneurs.

CEJA's biggest challenge is to reverse the ageing trend of the sector, in order to secure the future of European agriculture. Access to the agricultural sector is particularly difficult for young people, who face numerous challenges: access to land, access to credit, and high investments with low returns, and CEJA tries to raise the awareness of these issues among European decision-makers, calling for measures, such as installation aid, to facilitate entry to the sector and help buffer young farmers from market volatility and price fluctuations in the first few fragile years of business. CEJA attempts to achieve this by playing an important part in drafting the regulatory framework of agricultural policy at EU level.

CEJA also carries out projects. The Climate Farmers project, conducted by CEJA in collaboration with NAJK (Dutch Agricultural Youth Organisation), compiled the best practices of young farmers reducing greenhouse gas (GHG) emissions on their farms, and in 2011 published a booklet documenting 12 best practices. The Farm and Farmers project developed videos showing how young European farmers are finding innovative ways to support themselves in a changing farming environment.[74] The European School Competition project motivates pupils to create their own advertising campaign to encourage young people to become farmers or to work in the agricultural sector.[75] Finally, CEJA's Future Food Farmers project is a European-wide pledge campaign which aims to raise public and political awareness of the impending age crisis in Europe through the use of pledges and promotional videos.

(Case study drafted by C. Goemans, adapted from Survey)

Information provided by Ms Pascale Rouhier, CEJA Secretary-General and Natalia Skupska, CEJA Project Officer. More information available at: www.ceja.org http://europa.eu/ http://ec.europa.eu/agriculture/cap-post-2013/

## 47. YPARD – YOUNG PROFESSIONALS' PLATFORM FOR AGRICULTURAL RESEARCH FOR DEVELOPMENT

One of YPARD's key objectives is: "broadening opportunities for young professionals to contribute to strategic agricultural research for development policy debates". Young people are included in these discussions, ensuring that the voice of youth is effectively communicated and audible.

YPARD mobilizes its members to become active in discussions on a global, regional, national and subnational level, such as GFAR (Global Forum on Agricultural Research) and its stakeholder groups.[76] It often approaches conference organizers to advocate a greater number of young professionals in the discussion and encourages sponsors to back young professionals who are rarely selected by their own institutions for attendance at conferences.

---

[74]  http://www.ceja.eu/en/projects/ceja-projects/farm-and-farmers-project
[75]  http://www.ceja.eu/en/projects/ceja-projects/european-school-competition-project
[76]  http://www.egfar.org/about-us/forums-stakeholders/regional-fora

Youth interventions are often born through wide-scale e-discussions, from which the most active are selected to participate in conferences. As a multistakeholder network in agricultural development, YPARD has contributed to various discussion topics, including: youth perspectives in extension and advisory services; regional youth perspectives in the Near East and North Africa;[77] innovations on extension and advisory services; and the global landscapes forum (prelude to the COP18[78],[79]).

Youth were in the spotlight at the second Global Conference on Agricultural Research for Development (GCARD2).[80] YPARD collaborated with CGIAR to bring 35 young social reporters to GCARD2. A global call was published on ypard.net and disseminated to more than 20 000 people in three languages and an online group was built. The young reporters were prolific and committed, boosting the profile of the conference worldwide and interacting online with a global audience. A total of 136 social reporters from 44 countries were involved in writing 63 mainstream articles, 152 blog posts, 1 500 daily Twitter updates and 300 Facebook updates. The group has continued to be active in agricultural development events globally, to ensure that youth and other marginalized stakeholders are represented in 2013 and beyond.[81] After the success of the young social media group at GCARD2, active young social media teams are increasingly sought out by agriculture-related conferences.

YPARD also targets youth involvement in agricultural policy by advocating for young people on executive and steering committees of influential organizations in ARD (agricultural and rural development). The "youth advisor" is able to exert influence and inform decision-making, providing a stronger link to young people in agriculture. It also normalizes the exchange between senior and junior colleagues, giving value to what young people have to say.

(Case study drafted by C. Paisley, YPARD Coordinator)

Information provided by Ms Courtney Paisley, YPARD Global Coordinator.
More information available at:
www.ypard.net

# 6.3 Conclusions

Through the International Year of Youth in 2011 [Introduction], youth have received attention and been endorsed as important partners for addressing the challenges resulting from the recent crises. In the words of the Secretary-General of the United Nations, Ban Ki-moon: Youth should be given a chance to take an active part in the decision-making of local, national and global levels.

Nevertheless, in the agricultural sector, youth have yet to have a full role in policy dialogue. In developing countries in particular, the inclusion of youth in agricultural policy-making is still in an exploratory stage and the full impact is yet to be felt. It is important to outline the processes, as FANRPAN did when conducting case studies on youth policies and initiatives in six African

---

[77]   http://www.arimnet.net/
[78]   http://lac.ypard.net/event/cop-18-doha-2012-united-nations-climate-change-conference
[79]   http://ypard.net/2012-october-9/asia-pacific-graduates%E2%80%99-youth-declaration-cop-18
[80]   http://www.egfar.org/gcard-2012
[81]   http://gcardblog.wordpress.com/2012/11/26/gcard2-making-agriculture-cool-again/

countries [case 42]. For an increased understanding of youth's challenges in the agricultural sector – and the reflection of these challenges in policies – data should be aggregated concerning age, sex and geographical location, and the aspirations, needs and concerns of young people as a heterogeneous group should be taken into account in order to come up with policies that make agriculture more attractive to them.

The examples presented show that it is crucial for rural youth to organize themselves or join an existing organization, providing a sustainable channel to get their voices heard and to actively engage in policy dialogue. The International Cooperative Youth Statement 2012, presented during the closing ceremonies of the International Year of Cooperatives, acknowledges the importance of cooperatives as "a viable and promising option for young peoples' transition to full economic, social and civic participation" and recommends authentic inclusion of youth in the governance of cooperatives. Youth-only organizations as well as youth sections within existing "mixed" organizations have been established at local and national level, and youth are also starting to get organized at regional level. For example, in Togo [case 40] and Nepal [case 43], effective youth structures were created within farmers' organizations at national level. In both cases, youth are represented in the organization's decision-making organs and form strong lobby groups. Special measures should be taken to facilitate young women's participation in organizations and to enhance their leadership skills. These measures could include: reducing women's workloads; building their capacities; setting young women quotas in membership and leadership of the organization; and sensitizing local leaders about the importance of young women's participation (World Bank/FAO/IFAD, 2009).

Advocacy activities should address the challenges faced by youth when they are trying to establish themselves in farming. In this regard, access to land remains one of the most important challenges. Young Filipinos united to call for more efficient implementation of the existing land laws [case 10], and Burkinabe young women approached their traditional authorities for the release of land [case 11]. In Togo, young farmers participated in the drafting of the national policy on access to land for youth and women through REJEPPAT [case 40], while at global level, YPARD lobbies to raise the voice of young professionals in agricultural development in high-level policy dialogue in ARD [case 47].

At regional level, CEJA has represented young farmers for many years [case 46], while the AU turned its attention to youth more recently, initiating a dialogue within the framework of its 2011 summit and the CAADP Partnership Platform Meeting in 2012 [case 41]. Apex organizations of farmers at regional level, such as the Subregional Platform of Peasant Organizations of Central Africa (PROPAC) have also begun to incorporate youth structures.[82] The Internet plays a vital role in connecting youth from various countries. For example, Facebook pages have been launched by the Pacific Youth in Agriculture Network[83] [case 44] and by national sections of CAFY[84], which encourages youth to discuss policies and strategies. Meanwhile, the global movement YPARD has made extensive use of various social networks to reach out to thousands of youths [case 47]. In 2011, MIJARC, FAO and IFAD launched a process of regional youth

---

[82]  The creation of national youth colleges and a regional youth college have been foreseen in the statutes of PROPAC with the objective of achieving a 20 percent youth representation within the organization. To date, 5 out of 14 PROPAC member organizations (in Cameroon, Congo, Democratic Republic of the Congo, Chad and Central African Republic) inserted such a structure. Source: Ms Marie Josée Ninon Medzeme Engama, member of PROPAC Cameroon.

[83]  http://www.facebook.com/#!/pages/Pacific-Youth-in-Agriculture-Network/124321184278078?fref=ts

[84]  CAFY was set up in 2002 as an advocacy group seeking to bring together youth stakeholders to encourage collective action and decision-making strategies to promote increased youth involvement in sustainable agriculture around the Caribbean region (see http://www.facebook.com/#!/pages/ST-Lucia-Agriculture-Forum-for-Youth/128448173843298?fref=ts).

consultations specifically related to the agricultural sector and the challenges faced by rural youth. Regional recommendations were drafted and then further fine-tuned and consolidated by the rural youth representatives during a preparatory session of the 2012 Farmers' Forum and ultimately presented at the IFAD Governing Council. Youth also participated in the high-level Civil Society Organizations (CSOs) forum held in parallel to the 2009 World Food Summit.[85]

The momentum of the International Year of Youth should be sustained. Coordination is needed between different ministries (ministry of youth, agriculture, labour etc.) at central and local levels to ensure that policies reflect youth and their engagement in agriculture. Civil society actors and the youth themselves should actively participate in the policy-making process so that policies are more sustainable and owned by the target group (UNESCO, 2004). As initial steps are taken to include youth in agricultural policy dialogue, efforts should be made to make youth participation systematic. Rural youth representatives should be invited to agricultural meetings and events on a regular basis. In Togo [case 40] it is clear that the national authorities are not always aware of youth issues. Leadership and perseverance are vital for establishing sustainable relations between young farmers and state institutions, a successful example of which is CEJA, which has provided a platform for dialogue between young farmers and European decision-makers since 1958 [case 46]. Brazil has also institutionalized dialogues on international agreements related to agriculture with youth at different levels of the society [case 45].

---

[85] The CSO forum that was held in parallel to the 2009 World Food Summit paid special attention to youth and 15 percent of the delegates that participated in this forum were youth. In the framework of this forum, a youth caucus was created and a youth forum was held to ensure active youth participation. The youth caucus drafted a declaration which was presented in the plenary of the CSO forum. The final statement of the CSO forum was presented by a young MIJARC representative in the plenary of the World Food Summit (International Planning Committee for Food Sovereignty, 2009; 2010; MIJARC, 2009).

# 7. Overall conclusions

While the last decade has seen accelerated economic growth in many developing countries (UNDP, 2013), the resulting income gains have often been inequitably distributed. As a result, this growth has not resulted in commensurate poverty reduction for the most vulnerable groups, particularly rural youth. Instead, they continue to suffer from disproportionately high levels of unemployment, underemployment and poverty. In 2012, close to 75 million young people worldwide were out of work, resulting in a global youth unemployment rate almost three times the corresponding rate for adults (ILO, 2013). Furthermore, among those young people who were working, over 200 million were earning less than USD 2 per day (ILO, 2012c). In Africa, the proportion of working youth earning less than USD 2 per day is over 70 percent (OECD, 2012), many of whom were living in the continent's economically stagnant rural areas

Compounding this problem is the size of the world's youth cohort. In 2010, 1.2 billion people (15 percent of the world population) were between 15 and 24 years of age, many of whom were living in developing countries. Asia alone is home to 60 percent of the world's youth, while a further 18 percent live in Africa. Within Africa, more than 61 percent of the entire population is under 25 years of age, representing current and future youth cohorts of a daunting magnitude. Rather than dissipate over time, the pressures on policymakers to find employment opportunities for these young people will only grow if left unaddressed. Failure to generate opportunity for the world's youth cohort – particularly those living in developing countries' economically stagnant rural areas – will undermine both current and future poverty reduction efforts.

In many countries, the agriculture sector possesses significant development potential which, if seized, could generate ample decent and gainful employment opportunities for rural youth. However, it is not only the agricultural sector that possesses untapped potential, but also the youth themselves. Their capacities for creativity and economic innovation are squandered when they are blocked from actively participating in economic activities. As a result, facilitating and incentivizing youth participation in the agriculture sector would not only provide much needed employment opportunities for youth themselves, but may also help drive the innovation and growth needed to reduce rural poverty among youths and adults alike. Unfortunately, many young people do not perceive agriculture as a viable or attractive means of earning a living. The drudgery of low-productivity agriculture is simply not attractive to youth, who instead migrate to cities in search of higher productivity and better-remunerated employment. A concerted and coordinated effort is therefore needed to develop a more modern agricultural sector in many developing countries, and thereby unlock the potential of the youth cohort.

Since the 2007–08 food crisis, policy makers have shown greater awareness of the challenges faced by youth,[86] and have refocused their attention on the agricultural sector (FAO, 2011c). This provides momentum for developing policies, programmes and projects to enable youth's gainful employment in agriculture. This publication has provided case study evidence of some of the options to seize the opportunities for engaging youth in agricultural activities, building on the comparative advantages youth possess, such as their flexibility and commitment. Government officials, policy-makers and development practitioners may take these cases into consideration when programming national youth development activities.

Most of the challenges analysed are interwoven, as are the solutions. Although youth are generally better educated than their parents' generation, they often lack the relevant knowledge and experience on matters ranging from access to land, financial services, green jobs and markets to

---

[86]  The UN General Assembly proclaimed 2010 as the International Year of Youth UN, 2010). The theme of the UN International Day of Cooperatives 2011 was "Youth, the future of cooperative enterprise" (available at http://www.copac.coop/idc/2011/index.html).

information on relevant policies. The youth informants who participated in the MIJARC/IFAD/ FAO project rated access to land and access to finance as the main challenges for starting an agricultural activity. Youth rarely own the assets (e.g. land) that are required as collateral to obtain loans, and they are often perceived by financial institutions as a high-risk category. Financial services specifically designed for youth are rare and an enabling regulatory environment is required to facilitate youth's access to financial services. Delayed inheritance, increased land fragmentation and degradation make it difficult for youth to access land. Youth do not normally have the required savings to buy land, and obtaining loans is difficult. Accessing markets for agricultural products can be particularly challenging for young people: they often lack the capacity to produce large quantities to benefit from economies of scale, they frequently lack the required knowledge of prices and market structures and have limited bargaining power. Rural youth rarely have an active role in the drafting of policies affecting them, as in rural communities decision-making is often perceived as the domain of older men. As a result, policies tend not to reflect rural youth's real needs. Rural youth organizations, where they exist, are young, small, without financial means and therefore unable to impact at a high level. Meanwhile, in mixed organizations,[87] rural youth rarely hold decision-making positions. Young rural women are often even more disadvantaged: cultural practices may mean that girls' freedom of movement is restricted and they are often expected to help with domestic chores, and they therefore have fewer opportunities to attend learning activities, participate in the activities of organizations and access markets. Furthermore, in many communities, regulations and tradition still prescribe that boys inherit land and that women can only access land through their relation with a male relative.

Rural youth encounter numerous challenges when starting agricultural activities, but they can be overcome. While the examples herein are context-specific, general conclusions can still be drawn.

>   Access to the right information can help overcome young farmers' lack of experience.

>   Integrated training approaches are required for youth so that they may respond to the needs of a modern agricultural sector.

>   Modern ICTs offer great potential: they can attract youth to the sector, provide up-to-date information, and are excellent marketing, training and financing tools.

>   Bringing youth together creates numerous opportunities, as rural organizations can be instrumental in achieving: economies of scale when buying agricultural inputs and selling agricultural products; access to financial services, as the group can serve as guarantor for its members, giving youth a lower risk profile; access to land, as youth can pool their resources to buy or lease land; and participation in policy-making.

>   Youth-specific projects and programmes, while not always ideal as youth would rather be recognized as a specific target group within general projects, can provide the extra push to enter the agricultural sector. For example, scholarships can facilitate higher agricultural education or, when offering loans to young people, a parallel financial management training programme can ensure that youth gain the necessary skills to pay back the instalments in time. Projects and programmes can build on youth's comparative advantages, such as their special interest in the conservation and management of natural resources; their eagerness to work with ICTs; and their creativity in exploring niche markets.

---

[87]   Mixed organizations are organizations comprising both young and older members.

> A coherent and integrated response is required to address the core challenges faced by youth when entering the agriculture sector. It is important to work in partnership, with a transparent multistakeholder mechanism ensuring coherence, coordination and cooperation across different national government institutions and agencies, at central and local level, private sector organizations, youth organizations and development partners.

A coordinated response to increase youth's access to the agricultural sector is more important than ever, as a rising global population and decreasing agricultural productivity gains imply that young people must play a pivotal role in ensuring a food-secure future for themselves, and for future generations.

# References

**AFDB/OECD/UNDP/UNECA (African Development Bank, Organisation for Economic Co-operation and Development, United Nations Development Programme, United Nations Economic Commission for Africa).** 2012. *African economic outlook 2012* (available at http://www.oecd.org/inclusive-growth/African%20Economic%20Outlook%202012.pdf).

**Atkinson, A. & Messy, F.** 2012. Measuring financial literacy: Results of the OECD/INFE (International Network on Financial Education) Pilot Study, *OECD working papers on finance, insurance and private pensions*, No. 15. OECD Publishing (available at http://dx.doi.org/10.1787/5k9csfs90fr4-en).

**Beintema, N.M. & Di Marcantonio, F.** 2009. *Women's participation in agricultural research and higher education: key trends in sub-Saharan Africa.* ASTI.

**Beintema, N.M. & Di Marcantonio, F.** 2010. *Female participation in African agricultural research and higher education: new insights synthesis of the ASTI-Award Benchmarking Survey on Gender-Disaggregated Capacity Indicators.* IFPRI Discussion Paper 00957.

**Bennell, P.** 2007. *Promoting livelihood opportunities for rural youth.* IFAD (available at http://www.ifad.org/events/gc/30/roundtable/youth/benell.pdf).

**Bennell, P.** 2010. *Investing in the future, creating opportunities for young rural people.* IFAD (available at http://www.youtheconomicopportunities.org/sites/default/files/uploads/resource/invest_future_IFAD.pdf).

**Blackie, M., Blackie, R., Lele, U. & Beintema, N.** 2010. *Capacity development and investment in agricultural R&D in Africa.* Lead background paper ministerial conference on higher education in agriculture in Africa. Speke Resort Hotel, Munyonyo, Kampala, Uganda, 15–19 November 2010.

**Dalla Valle, F.** 2012. *Exploring opportunities and constraints for young agro entrepreneurs in Africa.* Conference abridged version. Rome, FAO.

**DeMuth, S.** 1993. *Community supported agriculture (CSA): an annotated biography and resource guide.* Washington, DC, USDA (available at http://www.nal.usda.gov/afsic/pubs/csa/at93-02.shtml).

**Dirven, M.** 2010. *Juventudes rurales en América Latina Hoy: fortalezas y desafíos, con acento en el empleo.* Documento presentado en el Taller internacional "Jóvenes: protagonistas del desarrollo en los territorios rurales" organizado por el ministerio de Agricultura y Desarrollo rural de Colombia, el Proyecto Oportunidades Rurales, Corporación PROCASUR, FAO, FIDA y Fundación Ford en Bogotá, Colombia, entre el 27 el 29 de octubre 2010.

**EU (European Union).** 2012. *Women in EU agriculture and rural areas: hard work, low profile.* Agricultural Economic Brief No. 7.

**EU.** 2010. *Rural development in the European Union: statistical and economic information report.* Rome.

**FAO (Food and Agriculture Organization of the United Nations).** 1997. *Higher agricultural education and opportunities in rural development for women – An overview and summary of five case studies.* M. Karl. Rome.

**FAO.** 2005. *Rural-urban market linkages: an infrastructure identification and survey guide.* J. Tracy-White. Rome.

**FAO.** 2007. *Education for rural people and food security: a cross-country analysis,* P. De Muro & F. Burchi. Rome.

**FAO.** 2009a. *Education for rural people: the role of education, training and capacity development in poverty reduction and food security,* D. Acker, D. & L. Gasperini. Rome.

**FAO.** 2009b. *Linking people, places and products. A guide for promoting quality linked to geographical origin and sustainable geographical indications* (available at http://www.fao.org/docrep/012/i1057e/i1057e00.pdf).

**FAO.** 2010. *Payments for environmental services within the context of the green economy* (available at http://www.fao.org/docrep/013/al922e/al922e00.pdf).

**FAO.** 2011a. *The State of Food and Agriculture 2010–11: women in agriculture. Closing the gender gap for development*. Rome.

**FAO.** 2011b. *The state of the world's land and water resources for food and agriculture. Managing systems at risk*. Rome.

**FAO.** 2011c. *Food and agricultural policy trends after the 2008 food security crisis: renewed attention to agricultural development*, by M. Aguirre, S. Kim, M. Maetz, Y. Matinroshan, G. Pangrazio, & V. Pernechele V. 2011. Rome.

**FAO.** 2012. *Voluntary Guidelines on the Responsible Governance of Tenure of Land, Fisheries and Forests in the Context of National Food Security*. Rome.

**FAO/UNESCO.** 2003 . Education for rural development: towards new policy responses. D. Atchoarena & L. Gasperini. Rome.

**Fréguin-Gresh, S., Losch, B. & White, E.** 2010. *Structural dimensions of liberalization on agriculture and rural development: a cross-regional analysis on rural change*.

**Hartl, M.** 2009. *Technical and vocational education and training (TVET) and skills development for poverty reduction – do rural women benefit?* Paper presented at the FAO-IFAD-ILO Workshop on Gaps, trends and current research in gender dimensions of agricultural and rural employment: differentiated pathways out of poverty Rome, 31 March - 2 April 2009.

**Hellin, J., Lundi, M. & Meijer, M.** 2009. Farmer organization, collective action and market access in Meso America. *Food Policy*, 34(1): 16–22.

**IEG.** 2013. *World Bank and IFC Support for Youth Employment Programs*. Washington, DC, World Bank. DOI: 10.1596/978-0-8213-9794-7. License: Creative Commons Attribution CC BY 3.0.

**IFAD (International Fund for Agricultural Development).** 2004. *Trade and rural development: opportunities and challenges for the rural poor*. Discussion paper. Rome.

**IFAD.** 2010a. *Rural poverty report 2011: New realities, new challenges: new opportunities for tomorrow's generation*. Overview. Rome.

**IFAD.** 2010b. *Rwanda:Gender and youth in the tea and coffee value chains*, Smallholder Cash and Export Crops Development Project (PDCRE). Rome.

**IFAD.** 2010c. *IFAD decision tools for rural finance*. IFAD. Rome (available at http://www.ifad.org/ruralfinance/dt/full/dt_e_web.pdf).

**IFAD.** 2011. *Making a difference in Asia and the Pacific: creating opportunities for rural youth*. Newsletter Issue 35, Jan.–Feb. Rome.

**ILO (International Labour Organization).** 2009. *Skills for green jobs: A global view*. Geneva (available at http://www.ilo.org/skills/projects/WCMS_115959/lang--en/index.htm).

**ILO.** 2011. *Assessing green jobs potential in developing countries: a practitioner's guide*. Geneva.

**ILO.** 2012a. *Promoting green entrepreneurship: first lessons from the Youth Entrepreneurship Facility in Kenya 2010–2011*. Geneva (available at http://www.ilo.org/wcmsp5/groups/public/---ed_emp/---emp_ent/documents/publication/wcms_176540.pdf).

**ILO.** 2012b. *Global employment trends for youth 2013, A generation at risk*. Geneva.

**ILO.** 2012c. *The youth employment crisis: a call for action*. Resolution and conclusions of the 101st Session of the International Labour Conference. Geneva.

**ILO.** 2013. *ILO global employment trends for youth 2013*. Geneva.

**ILO/UNEP.** 2012. *Working towards sustainable development. Opportunities for decent work and social inclusion in a green economy*. Geneva.

**International Planning Committee for Food Sovereignty.** 2010. *Final report of the People's Food Sovereignty Forum*, 13–17 November 2009, Rome.

**International Planning Committee for Food Sovereignty.** 2009. Food Sovereignty now: Young people creating their future. People's Food Sovereignty Forum 13–17 Nov. 2009, Rome.

**ITU (International Telecommunication Union).** 2010. *Monitoring the WSIS targets: a mid-term review*. World telecommunication/ICT development report 2010.

**Kruijssen, F., Keiizer, M. & Giuliani, A.** 2009. Collective action for small-scale producers of agricultural biodiversity products. *Food Policy* 34(1): 16–52.

**Leavy, J. & Smith, S.** 2010. Future farmers? Exploring youth aspirations for African agriculture. *Future Agricultures*, Policy Brief, No. 037.

**Lintelo, D. te.** 2011. *Youth and policy processes.* Working paper, Future Agricultures.

**Llanto, G.M. & Ballesteros, M.M.** 2003. *Land issues in poverty reduction strategies and the development agenda: the Philippines.* In *FAO/WB Land reform: land settlement and cooperatives.*

**McGregor, A., Tora, L., Bamford, G. & McGregor, K.** 2011. *The Tutu Rural Training Centre: lessons in non-formal adult education for self-employed in agriculture.*

**MIJARC (Mouvement international de la jeunesse agricole et rurale catholique).** 2009. *Annual Report 2009 of MIJARC/IMCARY.*

**MIJARC/IFAD/FAO.** 2012. *Summary of the findings of the project implemented by MIJARC in collaboration with FAO and IFAD: 'Facilitating access of rural youth to agricultural activities'.* The Farmers' Forum Youth session, 18 February 2012 (available at http://www.ifad.org/farmer/2012/youth/report.pdf).

**Ncube, M. & Anjanwu, J.C.** 2012. Inequality and Arab spring revolutions in North Africa and the Middle East. *Africa Economic Brief*, 3(7).

**OECD (Organisation for Economic Cooperation and Development).** 2012. *African Economic Outlook 2012.* Paris.

**PAFPNet (Pacific Agricultural and Forestry Policy Network).** 2010. *Youth in agriculture strategy 2011–2015: echoing the voices of Pacific youth.* Compiled by Pacific Agricultural and Forestry Policy Network, Secretariat of the Pacifi c Community Land Resources Division.

**Paisley, C.** 2012. *Skill gaps in formal higher agricultural education: a youth perspective.* Background paper for the Future Agricultures Conference on Young People, Farming and Food: The Future of the Agrifood Sector in Africa.

**Proctor, F. & Lucchesi, V.** 2012. *Small-scale farming and youth in an era of rapid rural change.* London, IIED.

**Shrader, L., Kamal, N., Darmono, W.A. & Johnston, D.** 2006. *Youth and Access to Microfinance in Indonesia: Outreach and Options.* A study jointly commissioned and financed by ImagineNations Group and The World Bank, co-founding partners of the Global Partnership for Youth Investment. Jakarta.

**The Youth Employment Network.** Undated. *Joining forces with young people: a practical guide to collaboration for youth employment* (available at http://www.ilo.org/public/english/employment/yen/downloads/yen_youth_guide_en.pdf).

**UN (United Nations).** 2009. *Millennium development goals report.* New York.

**UN.** 2010. *Resolution adopted by the General Assembly on 18 December 2009 A/RES/64/134* (available at http://www.un.org/en/ga/search/view_doc.asp?symbol=A/RES/64/134).

**UNESCO (United Nations Educational, Scientific and Cultural Organization).** 2004. *Empowering youth through national policies.*

**UNESCO.** 2010. *Reaching the marginalized.* Education for all global monitoring report. EFA Global Monitoring Report.

**UNESCO.** 2012. *Reaching out-of-school children is crucial for development.* UIS Factsheet, June 2012, No. 18.

**UNDESA (United Nations, Department of Economic and Social Affairs, Population Division).** 2010. *World youth report: youth and climate change.* New York.

**UNDESA.** 2010b. *World Programme of Action for Youth.* New York.

**UNDESA.** 2011. *World population prospects: the 2010 revision, highlights and advance tables.* Working Paper No. ESA/WP. 220.

**UNDESA.** 2012. *World Youth Report 2011: youth employment, youth perspectives on the pursuit of decent work in changing times.* New York.

**UNCDF (UN Capital Development Fund).** 2012. *Policy opportunities and constraints to access youth financial services.* Insights from the UNCDF's Youthstart programme (available at http://www.uncdf.org/sites/default/files/Download/AccesstoYFS_05_for_printing.pdf).

**UNDP (United Nations Development Programme).** 2013. Human development report 2013. New York.

**UNEP (United Nations Environment Programme).** 2008. Green jobs: Towards decent work in a sustainable, low-carbon world. Report commissioned and funded by UNEP, as part of the joint UNEP, ILO, IOE, ITUC Green Jobs Initiative.

**UNEP.** 2010. *Green economy. Developing countries success stories.* Geneva.

**UN-HABITAT (United Nations Human Settlements Programme).** 2011. *Towards a youth agenda for the global land tool network: a scoping study*. UN-HABITAT, Kenya.

**USAID.** 2005. *Enhancing women's market access and promoting pro-poor growth.*

**White, B.** 2012. *Agriculture and the generation problem: rural youth, employment and the future of farming.* Paper for the FAC–ISSER Conference on Young People, Farming and Food, Accra, 19–21 March 2012.

**World Bank.** 2008. *Teachers for rural schools: Experiences in Lesotho, Malawi, Mozambique, Tanzania and Uganda*. A. Mulkeen & D. Chen. Africa Human Development Series. WB.

**World Bank.** 2009. *Africa development indicators 2008/2009. Youth and employment in Africa: the potential, the problem, the promise.* Washington DC.

**World Bank.** 2011a. *World development report 2012: gender equality and development*. Washington, DC, International Bank for Reconstruction and Development/World Bank.

**World Bank.** 2011b. *ICT in agriculture, connecting smallholders to knowledge, network and institutions.* E-sourcebook on ICT in agriculture. Report Number 64605. Washington, DC, International Bank for Reconstruction and Development/World Bank.

**World Bank/FAO/IFAD.** 2009. *Gender in agriculture sourcebook.* Washington, DC, International Bank for Reconstruction and Development/World Bank.

**World Bank/UNICEF/USAID.** 2009. *Youth socio-economic empowerment through Inclusive Business Development and Innovative Social Service Delivery Project.*

**van Schalkwyk, H.D., Groenewald, J.A., Fraser, G.C.G., Obi, A. & van Tilburg, A.** 2012. *Unlocking markets to smallholders.* Wageningen Academic Publishers.

# Annex I: Survey 'Youth and agriculture: key challenges and concrete solutions'

## 1. INFORMATION REGARDING THE RESPONDENT

**1. Please provide your contact details:**

> Name ...............................................................................................................

> Surname ...........................................................................................................

> Address ............................................................................................................

> ........................................................................................................................

> Country ............................................................................................................

> Phone no. .........................................................................................................

> E-mail ..............................................................................................................

**2. You are**
☐ A man
☐ A woman

**3. Please provide your job title**
> ........................................................................................................................

> ........................................................................................................................

**4. If you are filling in this questionnaire on behalf of, or in relation to an organization, please provide name and contact details of the organization(s)**

> Name of organization .......................................................................................

> Address ............................................................................................................

> Country ............................................................................................................

> Phone no. .........................................................................................................

> E-mail ..............................................................................................................

> Web site ...........................................................................................................

**5. Is this organization registered?**

☐ Yes

☐ No

**6. Indicate the type of this organization. Please only tick the type(s) that best fit(s) your organization's profile (you may tick more than one box).**

☐ Producers' organization (producers include women and men, youth and elderly who are involved in agriculture, fisheries, livestock and forestry sectors)

☐ Young producers' organization (with only young members)

☐ Women producers' organization (with only women members)

☐ Youth association

☐ Women's association

☐ Governmental institution

☐ Non-governmental Organization

☐ Intergovernmental Organization

☐ Donor organization

☐ Research institution

☐ Private company

☐ Development organization

## 2. INFORMATION REGARDING THE CHALLENGE

**7. Which of the challenge(s) below does your solution address? Please only tick the most relevant boxes (you may tick more than one box).**

☐ Limited or no access to land (including inheritance issues; access to finance to purchase land; land grabbing; customs and traditions that hinder youth's access to land)

☐ Limited or no access to finance (including lack of access to savings; credit; and insurance)

☐ Limited or no access to input and output markets (including access to information on markets; capacity to counter other market actors; access to niche markets)

☐ Limited or no access to knowledge, skills and information (including agriculture in schools; vocational training; intergenerational knowledge sharing; leadership skills; ICTs; image of agriculture; financial literacy, business and entrepreneurship skills; business advisory services)

☐ Problematic intergenerational transfer of family farms and small-scale agricultural enterprises (including bureaucracy; installation aid; succession planning)

☐ Limited or no access to green jobs (including organic agriculture; ecotourism; employment related to using energy from renewable sources)

☐ Limited or no engagement of youth in policy dialogue (including youth organization; representation and leadership; space for dialogue)

☐ Other, please explain

> ........................................................................................................................................

........................................................................................................................................

........................................................................................................................................

8.  **Briefly describe the main characteristics of the chosen challenge(s) in your geographic area.**

> ...........................................................................................................................

...........................................................................................................................

...........................................................................................................................

9.  **Does the challenge affect male and female youth differently?**
☐  Yes
☐  No

10. **If yes, how?**

> ...........................................................................................................................

...........................................................................................................................

...........................................................................................................................

## 3. INFORMATION REGARDING THE SOLUTION

11. **Describe the solution that you are proposing to be included in the FAO publication to address the above selected challenge(s). Go into detail on how the solution works.**

> ...........................................................................................................................

...........................................................................................................................

...........................................................................................................................

...........................................................................................................................

12. **When was this solution adopted?**

> ...........................................................................................................................

13. **List the key success factors of the solution.**

> ...........................................................................................................................

...........................................................................................................................

...........................................................................................................................

**14. What were the difficulties/problems that had to be overcome to make the solution a success?**

> ...............................................................................................................................

.................................................................................................................................

.................................................................................................................................

**15. How have youth been involved in this solution? What has been their role in this solution?**

> ...............................................................................................................................

.................................................................................................................................

.................................................................................................................................

**16. Indicate the region(s) in which the solution has been used (you may tick more than one box).**
☐ Sub-Saharan Africa
☐ Latin America and the Caribbean
☐ Asia and the Pacific
☐ North Africa and Near East
☐ Europe and North America

**17. Specify the country/ies and exact locations in which the solution has been used.**

> ...............................................................................................................................

.................................................................................................................................

**18. Did youth organize themselves to overcome the challenge?**
☐ Yes
☐ No

**19. If yes, how were they organized and how did this possibly make a difference for addressing the challenge(s)?**

> ...............................................................................................................................

.................................................................................................................................

.................................................................................................................................

**20. Has the solution worked equally well for female and male youth?**
> If yes, please explain ...........................................................................................................

.................................................................................................................................

> If no, please explain ...................................................................................................................................

.......................................................................................................................................................................

**21. Are there specific gender-related differences that were considered in the solution? If yes, please explain.**

> ...............................................................................................................................................................

.......................................................................................................................................................................

.......................................................................................................................................................................

**22. Please provide an overview of the costs necessary to implement the solution. Please specify the currency.**

> ...............................................................................................................................................................

.......................................................................................................................................................................

.......................................................................................................................................................................

**23. List the conditions necessary for successful replication of the solution.**

> ...............................................................................................................................................................

.......................................................................................................................................................................

.......................................................................................................................................................................

**24. What were the results and impact of the solution on youth and their communities? If available, please also provide quantitative data.**

> ...............................................................................................................................................................

.......................................................................................................................................................................

.......................................................................................................................................................................

**25. Please list other institutions and organizations, partners, implementing agencies and donors involved in the solution and the nature of their involvement.**

> ...............................................................................................................................................................

.......................................................................................................................................................................

**26. Please list links of information resources referring to your solution and/or send information resources by e-mail.**

> ...............................................................................................................................................................